石油专业英语核心词汇一本通

全晓虎　范玉然　王同德　姚登樽　◎编著

江淑娟　◎主审

TECHNICAL PETROLEUM
ENGLISH VOCABULARY

石油工业出版社

内容提要

本书汇集了油气勘探开发、油气井工程、油气储运和石油炼化等专业核心技术英语词汇,所有词汇均辅以例句,使读者对相关词汇更精准理解。

本书可供石油石化行业技术与管理人员,以及高等院校相关专业的师生参考使用。

图书在版编目(CIP)数据

石油专业英语核心词汇一本通 / 全晓虎等编著.
—北京:石油工业出版社,2023.12
ISBN 978-7-5183-6505-0

Ⅰ.①石… Ⅱ.①全… Ⅲ.①石油工程–英语–词汇–自学参考资料 Ⅳ.①TE6

中国国家版本馆CIP数据核字(2023)第257325号

出版发行:石油工业出版社
　　　　　(北京市朝阳区安华里2区1号楼 100011)
　　　　网　　址:www.petropub.com
　　　　编 辑 部:(010)64523687　图书营销中心:(010)64523633
经　　销:全国新华书店
印　　刷:北京晨旭印刷厂

2023年12月第1版　　2023年12月第1次印刷
787×1092毫米　　开本:1/16　印张:21.75
字数:410千字

定　价:100.00元
(如出现印装质量问题,我社图书营销中心负责调换)
版权所有,翻印必究

前　言

随着世界能源步入低碳转型期，虽然新能源开发呈增长态势，但其分散性、不稳定性、成本高等弊端依然突出，传统化石能源仍将在一段时期内与可再生能源协同发展。在传统能源体系中，石油占据重要的份额，仍是保障能源安全的压舱石；天然气作为碳排放强度最低的化石能源，将成为能源转型中新能源的最佳伙伴。然而，我国原油和天然气对外依存度持续攀升，实现能源清洁高效利用任重道远。目前海外油气投资市场前景广阔，我们需要及时掌握国际核心技术，完成从常规向非常规油气、陆上向海上油气开发，以及从中浅层向深层、超深层开采的跨越，实现资源有序接替，以维护国家能源安全，成功完成低碳转型。

在此形式下，对既掌握石油专业知识又能无障碍进行国际交流的石油国际化人才的需求日益增长，同时也为各大高校即将毕业的莘莘学子提供了宝贵的就业机会。为了辅助业界选拔并储备国际化合作人才，相关行业资质考试也陆续推出。为了帮助非石油专业以及石油专业低年级学生奠定专业英语学习基础，应对石油石化行业相关考试，并获取该领域就业机会，本书作者基于多年石油专业英语教学经验、相关教材编写经验以及油田现场工作经验等，从词汇教学入手，编写了《石油专业英语核心词汇一本通》，以便能激发读者学习石油专业英语的热情和兴趣。

本书精心筛选了石油石化专业的英文核心词汇，覆盖了上、中、下游生产全链条，包括油气勘探开发、油气井工程、油气储运和石油炼化等所有环节。每部分词汇的排列是基于专业知识概念展开的，完整、系统、有序，每条词汇后标注词性和释义，并辅以例句，例句选择是基于专业相关性和基础可读性等原则，难易程度适中，并配有语言流畅、准确专业的中文译文，使读者能轻松快速地掌握抽象生涩的专业词汇，同时系统了解石油石化行业生产运营的基本流程，并反过来更好地理解记住词汇。希望本书能对石油石化行业相关人员有所裨益，为我国能源发展贡献微薄之力，同时也欢迎读者朋友对书中不足之处给予批评指正。

中国石油大学（北京） 全晓虎

2023 年 2 月

目 录

Part I Petroleum Geology and Exploration 石油地质勘探

Section I　　Petroleum Geological Structure 石油地质构造 2

Section II　　Petroleum Generation 油气生成 12

Section III　　Tectonic Action 构造作用 17

Section IV　　Oil and Gas Reservoir 油气藏类型 25

Section V　　Fundamental Exploration Methods 基本勘探方法 33

Section VI　　Seismic Exploration 地震勘探 46

Section VII　　Reservoir Evaluation 油气藏评价 62

Part II Oil and Gas Drilling Engineering 油气井工程

Section I　　Wells Categories and Operators 油气井类型及作业人员 72

Section II　　Drilling Equipment 钻井设备 80

Section III　　Drilling Fluid 钻井液 91

Section IV　　Rotary Drilling 旋转钻井 97

Section V　　Well Cementation and Completion 固井与完井 108

Section VI　　Well Logging and Remedial Work 测井与修井 119

Part III Oil and Gas Exploitation 油气开发

Section I　　Fundamentals of Petroleum Production 油气开采基本概念 126

| Section II | Conventional Oil and Gas Recovery 常规油气开发 | 129 |
| Section III | Unconventional Oil and Gas Extraction 非常规油气生产 | 150 |

Part IV Petroleum Storage and Transportation 油气储运

Section I	Oil-gas Gathering and Transportation 油气集输	164
Section II	Oil and Gas Storage 油气储存	174
Section III	Oil, Gas and Water Separation 油气水分离	189
Section IV	Natural Gas Processing 天然气处理	212
Section V	Wastewater Treatment 污水处理	226
Section VI	Pipeline Transportation 管道运输	231
Section VII	Pipeline Pressurization and Maintenance 管道增压与维护	245
Section VIII	Oil Tank Transportation 油罐油轮运输	253

Part V Petroleum Refining 石油炼化

Section I	Primary Oil Refining 石油初炼	262
Section II	Secondary Oil Refining 石油精炼	276
Section III	Petroleum Fuels 石油燃料	290
Section IV	Chemical Raw Materials 化工原料	296
Section V	Chemical Products 化工产品	306

参考文献 311

石油专业核心词汇汉语拼音索引 313

Part 1

Petroleum Geology and Exploration
石油地质勘探

Section I

Petroleum Geological Structure
石油地质构造

petroleum geology　石油地质学

例：The application of chemistry to the study of rocks (geochemistry) has many uses in petroleum geology.

将化学用于岩石研究（地球化学）在石油地质学中有许多用途。

field geological survey　*np.* 野外地质调查

例：The data systematical analysis and field geological survey support occurrence of favorable geological conditions for the formation of primary reservoirs in the Mid-late Proterozoic strata in the study area.

数据系统分析和野外地质调查表明，该研究地区中晚期原生代地层有利于形成原生油气藏。

petrology　*n.* 岩石学

例：Petrology is the branch of geology that deals with the origin, composition, structure, and alteration of rocks.

岩石学是地质学的一个分支，研究岩石的起源、成分、结构和演变。

stratigraphy　*n.* 地层学

例：Research of the stratigraphy and the petrology suggest the fact.

地层学和岩石学的研究证明了这一点。

sedimentary rock　*np.* 沉积岩

例：Sedimentary rocks—the most common rock type near the surface—are also the most common reservoirs for water because they contain the most space that can be filled with water.

沉积岩是近地表最常见的岩石类型，也是最常见的储水层，因为沉积岩可储水空间最多。

sandstones 砂岩

例：<u>Sandstones</u> and limestones, the most common types of reservoir rocks, are generally porous.

砂岩和灰岩一般都有孔隙，是最常见的储油岩类型。

tight sandstone *np.* 致密砂岩

例：Porous-type sandstone reservoir is shorter and shorter and gas reservoir in <u>tight sandstone</u> is more and more important.

由于孔渗性良好的砂岩油气藏越来越少，致密砂岩气藏近年来逐渐成为勘探和开发热点。

limestone *n.* 石灰岩

例：For more highly cemented materials and <u>limestones</u>, significant difference in total porosity and effective porosity values may occur.

对于胶结程度较高的物质和石灰岩，总孔隙度和有效孔隙度值可能差异显著。

carbonate rock *np.* 碳酸盐岩

例：Although <u>carbonate rocks</u> can be clastic in origin, they are more commonly formed through processes of precipitation or the activity of organisms such as coral and algae.

虽然碳酸盐岩可能是由碎屑产生的，但通常通过降水或珊瑚和藻类等生物活动形成。

dolomite *n.* 白云岩

例：Limestone, dolostone or <u>dolomite</u>, and chalk are carbonate rocks.

石灰岩、白云石或白云岩和白垩地层都是碳酸盐岩。

marine shale *np.* 海相页岩

例：Illite is a group of clay minerals formed during the alteration of silicate minerals such as mica and feldspar and commonly found in <u>marine shales</u>.

蒙脱石是一组在云母、长石等硅酸盐矿物蚀变过程中形成的黏土矿物，常见于海相页岩中。

metamorphic rock *np.* 变质岩

例：The composition, texture, and structure of metamorphic rocks are affected by a number of factors.

变质岩的成分、结构和构造受到许多因素的影响。

igneous rocks *np.* 火成岩

例：Igneous rocks are produced by cooling and crystallization of molten rock-making material called magma or lava.

火成岩是由称为岩浆的熔融造岩物质冷却结晶而成的。

basement rock *np.* 基岩

例：Basement rock can be igneous, metamorphic, and/or sedimentary rock.

基岩可以是火成岩、变质岩和/或沉积岩。

evaporite *n.* 蒸发岩

例：Anhydrite is a member of the evaporite group of minerals and the soft rock comprising anhydrite formed by precipitation of calcium sulfate from evaporation of seawater.

硬石膏是蒸发岩类矿物中的一员，是由海水蒸发形成的硫酸钙沉淀形成的软岩。

homocline *n.* 同斜岩

例：Sedimentary rocks dipping uniformly in one direction are known as a homocline. Although homoclines are common, they do not form gas and oil traps.

沿一个方向均匀倾斜的沉积岩称为同斜岩。虽然同斜岩很常见，但不能形成油气圈闭。

parent magma *np.* 母岩浆

例：Some ore bodies have no direct known relationship to a parent magma, and circulating ground water seems to have played a key role in their origin.

一些矿体与母岩浆没有直接的已知关系，循环地下水似乎在其成因中起到了关键作用。

fossil n. 化石

例：Fossils can be found in sedimentary beds.
化石可以在沉积层中找到。

microfossil n. 微体化石

例：Macrofossils (somewhat sparse) are present in the formation but nowhere near the quantity of the microfossils.
大化石（有些稀疏）存在于地层中，但远不及微体化石的数量。

lower metamorphic strata np. 浅变质岩地层

例：Lower metamorphic strata are well developed with complicated structure in the northeastern area.
东北地区浅变质岩地层发育良好，构造十分复杂。

outcrop n. 露头

例：Sedimentary rocks are tagged by the fossils which they contain, and scattered outcrops may be correlated on the basis of these labels.
沉积岩以其所含的化石标记，分散的露头就可以根据这些标记进行对比。

fault n. 断层

例：Faults are breaks in the rocks along which one side has moved relative to the other.
断层是岩石中一方相对于另一方移动的断裂结构。

strike-slip fault np. 走滑断层

例：Strike-slip fault, also called transcurrent fault, wrench fault, or lateral fault, in geology, is a fracture in the rocks of Earth's crust.
走滑断层，在地质学中也称为跨流断层、扭断层或横向断层，是地壳岩石中的一种断裂。

antithetic fault np. 反向断层

例：Stretching structural pattern principally consists of listric normal fault and tilt block, antithetic fault and rollover anticline, and "horst-graben" structure.

伸展构造模式主要包括铲状正断层与倾斜块体、反向断层与翻转背斜和"角地堑"构造。

faulted anticline np. 断块背斜

例：In all these areas there are also petroleum accumulations in anticlinal traps, and also in <u>faulted anticlines</u> in which the dominant trapping mechanism is difficult to discern.

在所有这些地区，有些油藏属于背斜圈闭，还有些属于断块背斜油藏。断块背斜油藏中主要的圈闭机制复杂，很难弄清楚。

anticline n. 背斜

例：<u>Anticlines</u>, but not synclines, form high areas in reservoir rocks and can be gas and oil traps.

背斜（而不是向斜）在储集岩中形成高位区，可以成为油气圈闭。

salt ridge anticline np. 盐脊背斜

例：The salt flow resulted in several kinds of structural traps mainly including <u>salt ridge anticline</u>, salt dome and salt edge syncline.

盐体塑性流动形成几种构造圈闭，主要有盐脊背斜、盐丘穹窿和盐边向斜等。

syncline n. 向斜

例：An anticline is a large, upward arch of sedimentary rocks, whereas a <u>syncline</u> is a large, downward arch of rocks.

背斜是巨大的向上延伸的沉积岩拱，而向斜是巨大的向下延伸的岩石拱。

dome n. 穹窿

例：A <u>dome</u> is a circular or elliptical uplift. Domes also form gas and oil traps.

穹窿是圆形或椭圆形隆起物，还可以形成油气圈闭。

natural fracture np. 天然裂缝

例：Two types of <u>natural fractures</u> in rocks are joints and faults.

岩石中的两种天然裂缝是节理和断层。

fractures / fissure *np.* 裂缝/缝隙

例：Porous rocks may sometimes also contain <u>fractures</u> or <u>fissures</u>, which will add to the oil-storing capacity of the reservoir.

孔隙性岩石有时也可能有裂缝或缝隙，可以增强油藏的储油能力。

matrix pores or fractures *np.* 基质孔缝

例：The pore spaces are mainly residual intergranular porosity, dissolution porosity in clastic rock reservoir, dissolution <u>matrix pores or fractures</u> in volcanic reservoirs.

碎屑岩储层中主要的储集空间为残余粒间孔、碎屑岩储层溶蚀孔、火山岩储层溶蚀基质孔缝。

pore *n.* 孔隙

例：When a significant fraction of the <u>pores</u> is interconnected so that fluids can pass through the rock, the rock is permeable.

当大量孔隙相互连通，流体能够通过岩石时，岩石才具有渗透性。

original pore/primary pore *np.* 原生孔隙

例：The <u>original pore</u> is developed in the deposition of the material.

原生孔隙是在物质沉积过程中自然形成的。

induced pore/secondary pore *np.* 次生孔隙

例：<u>Induced pore</u> is developed by some geological process subsequent to deposition of the rock.

次生孔隙是岩石沉积以后又诱发于某些地质作用而产生的孔隙。

heterogeneity *n.* 非均质性

例：The fractured hydrocarbon reservoir is featured in its complicated pores and natural fractures pattern, strong <u>heterogeneity</u>, and complex fluid property.

裂缝性油气藏孔隙—裂缝结构复杂，非均质性极强，流体性质难以确定。

joint/cleat *n.* 节理

例：A <u>joint</u> is a fracture in the rocks without any movement of one side relative to the other.

节理即岩层断裂后两侧未发生任何相对移动的裂缝结构。

lithology n. 岩性

例：<u>Lithology</u> identification is the base of reservoir study.

岩性识别是储层研究的基础和关键。

sorting n. 分选性

例：The lithology of reservoir is chiefly mid-fine grained sandstone with good <u>sorting</u> and low texture maturity.

该区块储层岩性以中细粒砂岩为主，分选性好，结构成熟度低。

throat n. 喉道

例：The <u>throat</u> characteristic of the low permeability reservoir is the crucial factor that decides if physical quality is good in low permeability layer.

低渗透储层喉道特征是决定储层物性好坏的关键因素。

sediment n. 沉积物

例：The rate of deposition of the <u>sediments</u> must be sufficiently rapid for the organic matter to be preserved by burial before being destroyed by decay.

沉积物的沉积速度必须足够快，以便有机物在腐烂之前掩埋保存下来。

sedimentary basin np. 沉积盆地

例：The gravity method has been extensively used in the location of <u>sedimentary basins</u>.

重力勘探法已广泛用于沉积盆地探测。

sedimentary structure np. 沉积构造

例：According to the <u>sedimentary structure</u> and lithology, the turbidite may be divided into the four types of turbidite facies.

按照沉积结构和岩性特征，可划分出四种浊积岩相类型。

depression n. 凹陷

例：The sedimentary basin is a <u>depression</u> filled with sedimentary rocks.

沉积盆地是指充填有沉积岩的凹陷。

fault basin *np.* 断陷盆地

例：Miyang Depression is a Mesozoic-Cenozoic fault basin, where lithologic change is fast, fault is developed, reservoir type is various.
泌阳坳陷为中新生代断陷盆地，岩性变化快，断裂发育，油藏类型多。

deposition system *np.* 沉积体系

例：The study on the deposition system would be the premise and base of the coal exploration.
对沉积体系的研究是煤炭勘探的前提与基础。

lacustrine facies *np.* 湖泊相

例：The main sedimentary facies in this area include lacustrine facies and gravity flow facies.
该区块沉积相以湖泊相和重力流相为主。

submarine canyon *np.* 海底峡谷

例：Submarine canyons are relatively common throughout the world and often occur offshore from rivers.
海底峡谷在世界各地都比较常见，而且经常在河流的近海附近出现。

fluid phase *np.* 流体相

例：Fluid phase equilibria among sour gas reservoirs have become major project and hot point of special gas reservoirs recently.
目前含硫气藏的流体相平衡已成为特殊气藏开发研究的重点和热点。

aqueous phase *np.* 水相

例：The invasion in solid phase and trap damage in aqueous phase are main reasons of productivity impairment.
固相侵入和水相圈闭损害是产能损失的主要原因。

gas phase *np.* 气相

例：Condensate is in a gas phase in the accumulation and in a liquid phase at the surface.
凝析油在地下处于气相，而在地表呈液相。

turbidity current *np.* 浊流

例：The sands in turbidity current deposits are called turbidites and can be reservoirs for gas and oil.

浊流沉积物中的砂层为浊积岩，可以称为油气储层。

salt structure *np.* 盐构造

例：About 60% petroliferous traps are associated with the salt structure in the world.

世界上大约有 60% 的含油气圈闭与盐构造有关。

formation sequence/stratigraphical sequence *np.* 地层层序

例：Paleo morphology controlled the formation sequence and development of continental basins, especially of rift basins.

古地貌对地层层序以及陆相盆地发育，尤其是裂谷盆地发育起着重要的作用。

sequence stratigraphy *np.* 层序地层学

例：Sequence stratigraphy can be used to understand the distribution of sandstone and shale in a particular area as a package of related sedimentary rock.

层序地层学可用于研究砂岩和页岩（作为一组相关沉积岩）在某区域内的分布情况。

aerated layer *np.* 风化层

例：An aerated layer typically has a low seismic velocity.

风化层通常具有较低的地震速度。

alluvium *n.* 冲积层

例：Alluvium is material deposited in an alluvial environment, typically detrital sediments that are poorly sorted.

冲积层是在冲积环境中沉积的物质，通常是分选不良的碎屑沉积物。

shale reservoir *np.* 页岩储层

例：Most horizontal wells in shale reservoirs are drilled in the direction of the minimum horizontal stress.

在页岩储层中，大多数水平井都是沿最小水平应力方向钻井。

biostratigraphic zones *np.* 生物层序地层带

例：Biology is applied to geology in several ways, notably through the study of fossils (paleontology), and is especially significant in establishing biostratigraphic zones for regional stratigraphical correlation.

生物学从多方面应用于地质学，尤其是用于化石研究（古生物学），为作区域地层对比而建立生物层序地层带具有极为重要的意义。

continental shelf *np.* 大陆架

例：Sedimentary rocks that are encountered in drilling along the beach extend out under the continental shelf.

沿着延伸至大陆架下方的海滩钻井时会发现沉积岩。

continental slope *np.* 大陆坡

例：Seaward of the continental shelf and slope break is the continental slope that extends down to the bottom of the ocean.

大陆架的向海斜坡和坡折带是向海洋底部延伸的大陆坡。

Section II

Petroleum Generation
油气生成

petroleum *n.* 石油

例：<u>Petroleum</u> is a result of the deposition of plant or animal matter in areas which are slowly subsiding.

石油是植物或动物等物质缓慢沉降于地下而产生的。

hydrocarbon *n.* 碳氢化合物

例：Petroleum is a highly variable mixture of <u>hydrocarbons</u> mixed with oxygen, sulphur, nitrogen, and other elements, including both liquids and gases.

石油是由氧、硫、氮和其他元素混合在一起的多种碳氢化合物，包括液体和气体。

crude oil *np.* 原油

例：<u>Crude oils</u> vary in color, from clear to tar-black, and in viscosity, from water to almost solid.

各种原油颜色不同，黏度也不同，从透明到焦油黑，从水样到几乎固态都有。

natural gas *np.* 天然气

例：<u>Natural gas</u> is a combination of hydrocarbon gases consisting primarily of methane (CH_4) and, to a lesser extent, butane, ethane, propane, and other gases.

天然气是碳氢化合物，主要成分是甲烷，另有少量的乙烷、丙烷、丁烷和其他气体。

organic origin *np.* 有机成因

例：According to petroleum <u>organic origin</u> theory, the original materials for hydrocarbon generation are the dispersed organic matters in rocks.

油气有机成因说认为，岩石中的分散有机质是生成油气的原始物质。

kerogen *n.* 干酪根

例：The remains of marine plant and animal life were deposited along with rock-forming sediments under the sea where they were decomposed anaerobically by bacteria into an asphaltic material called <u>kerogen</u>.

海洋动植物遗骸与海底成岩沉积物一起沉积，通过细菌厌氧分解转化为一种称为干酪根的沥青物质。

fossil fuel *np.* 化石燃料

例：Crude oil is known as a <u>fossil fuel</u> as it is derived from ancient organic material which existed many millions of years ago.

原油是从数百万年前古代有机物质中产生的，因此称为化石燃料。

inorganic origin *np.* 无机成因

例：According to <u>inorganic origin</u> theory, petroleum was formed by the chemical reaction of inorganic carbon and hydrogen in nature.

无机成因说认为，石油是由自然界的无机碳和氢经过化学作用而形成的。

generation *n.* 生成

例：Eventually chemical changes result in the <u>generation</u> of petroleum, a complex, highly variable mixture of hydrocarbons, including both liquids and gases.

最终经化学变化形成石油，一种复杂、多样的碳氢化合物的液体和气体混合物。

seep *n.* 油气苗

例：Some of the oil eventually reached the surface and collected in large pools of tar, which are called "<u>seeps</u>".

部分石油到达地面形成大型焦油池，称为"油气苗"。

oil and gas seepage *np.* 油气显示

例：The south of Zhungeer basin has complicated structure, many <u>oil and gas seepage</u> active surface indication, all this indicates a large resource quantity.

准噶尔盆地南缘构造复杂，圈闭众多，油气显示活跃，资源量大。

in situ *prep.* 原地

例：In certain cases, such as with coal and some carbonate deposits, the sedimentary material is formed in situ.

在某些情况下，沉积物质是在原地生成的，例如煤和一些碳酸盐岩沉积。

shale gas *np.* 页岩气

例：Shale gas is gas that has remained trapped in, or close to, its source rock.

页岩气是其源岩中或源岩附近的天然气。

tight gas *np.* 致密气

例：Tight gas are produced from a relatively impermeable reservoir rock.

致密气产自相对不渗透的储层岩石。

biogenic gas *np.* 生物气

例：During early diagenesis, microbial activity is a key contributor to the breakdown of organic matter and generally results in production of biogenic gas.

在成岩作用早期，微生物活动是有机质分解的关键因素，通常导致生物气的产生。

coalbed methane *np.* 煤层气

例：The coalbed methane content is the most important index for the evaluation of coalbed methane.

煤层甲烷含气量是评价煤层气的最重要的指标。

bituminous sand *np.* 沥青砂

例：Solid petroleum materials, such as natural bitumen, natural asphalt, and bituminous sands, are considered petroleum.

天然沥青和沥青砂这类固态石油物质被视为石油。

heavy tar/ tar mat *np.* 重油（焦油）垫

例：Some oil fields have a mat of heavy tar at the oil water contact. Tar mats cause considerable production problems by inhibiting water from displacing oil.

一些油田的油水界面处常形成一层重油垫。焦油垫会抑制水驱油作用，造成严重的生产问题。

unconventional oil and gas *np.* 非常规油气

例：In terms of the chemical composition, <u>unconventional oil and gas</u> are identical to conventional.

在化学成分上，非常规油气与常规油气是相同的。

conventional deposit *np.* 常规油气资源

例：Unconventional resources are less concentrated than <u>conventional deposits</u> and do not give themselves up easily.

与常规油气相比，非常规油气集中度较低，不容易开采。

capillary force *np.* 毛细管力

例：Main driving force for migration of separate phase natural gas is buoyancy, and resistance is <u>capillary force</u>.

在静水条件下，浮力是游离相天然气运移的主要动力，毛细管力是运移的阻力。

migrate *v.* 运移

例：Seeping through cracks and fissures, oozing through minute connections between the rock grains, all petroleum can <u>migrate</u> upward to the surface of the earth.

所有石油都可以通过裂缝以及岩石颗粒之间的微小通道渗出，向上运移到地球表面。

primary migration *np.* 初次运移

例：The elapsed time of hydrocarbons from deposition to <u>primary migration</u> was quite long.

烃类物质从沉积到初次运移要经过很长时间。

secondary migration *np.* 二次运移

例：The further movement of the hydrocarbons into reservoir rock in hydrocarbon trap or other area of accumulation is <u>secondary migration</u>.

二次运移指烃类继续运移进入储集岩石中圈闭，或进入其他聚集区。

migration path *np.* 运移路径

例：The major reservoir rock indicates the optimum migration path for the petroleum between the pod of active source rock and the traps that include the major reservoir rock.

主要储集岩具有自有效烃源岩至圈闭之间的最佳运移路径。

accumulation *n.* 成藏

例：Accumulation is the phase in the development of a petroleum system during which hydrocarbons migrate into and remain trapped in a reservoir.

成藏是油气系统发育的阶段，在此期间，油气运移到储层中并停滞。

reservoir forming assemblage *np.* 成藏组合

例：The reservoir forming assemblage is a series of hydrocarbon reservoirs and prospective traps with common history of hydrocarbon generation, migration and accumulation in a certain stratum.

成藏组合是某一地层段内具有共同油气生成、运移和聚集历史的一系列油气藏和潜在圈闭。

oil type gas *np.* 油型气

例：Shallow gas reservoir is that buried depth is less than 1,500 meters, including biogas, coal formed gas, oil type gas and water soluble gas etc.

浅层气藏是指埋藏深度小于1500米的气藏，主要包括生物气、煤成气、油型气和水溶气等。

shallow transitional gas *np.* 浅层过渡气

例：A lot of shallow transitional gas pools actually resulted from anaerobic degradation.

许多浅层过渡气实际上是由于原油厌氧菌降解而产生的。

Section III

Tectonic Action
构造作用

---

diagenesis *n.* 成岩作用

例：The reservoir is closely controlled by sedimentation and diagenesis.
储层主要受沉积作用和成岩作用的影响。

compaction *n.* 压实作用

例：Compaction and cementation of sediments to the degree that they become coherent, relatively solid rock.
沉积物的压实作用和胶结作用，使其成为紧密的、相对坚固的岩石。

folding *n.* 褶皱作用

例：When crustal forces act slowly, layered rocks under great pressure tend to be deformed by warping or folding.
当地壳力慢慢起作用时，层状岩石在巨大压力下容易因翘曲或褶皱作用而变形。

metamorphism *n.* 变质作用

例：Another type of metamorphism may be called contact metamorphism.
另一种变质作用可称为接触变质作用。

metamorphic process *np.* 变质过程

例：The metamorphic process has led to the development of a series of foliated rocks.
这个变质过程导致了一系列叶理化岩石发育。

cementation *n.* 胶结作用

例：Decrease of sandstone porosity is mainly caused by compaction and cementation.
压实作用和胶结作用是砂岩孔隙度降低的主要原因。

weathering n. 风化作用

例：Weathering of rock is essentially a static process.
岩石风化本质上是一个静态过程。

dissolution n. 溶蚀作用

例：The major diagenesis of reservoir is mechanical compaction, cementation, replacementation and dissolution.
储层的主要成岩作用为压实作用、胶结作用、置换作用和溶蚀作用。

tectonism n. 构造作用

例：Primary reasons leading to reservoir heterogeneity are sedimentary environment and tectonism.
引起储层非均质性的主要原因是沉积环境和构造作用。

tectonic stress np. 构造应力

例：Hydrocarbon migration and accumulation are the results of combined action of various factors, but tectonic stress is the main driving force of migration and accumulation during tectonic active period.
油气运移是各种因素综合作用的结果，但在构造活动期，构造应力是主要运移动力。

tectonic process np. 构造作用

例：The depression, formed by any tectonic process, is lined by basement rock.
构造作用形成的凹陷由基岩充填而成。

tectonic inversion np. 构造演化

例：This basin developed under extensional conditions and underwent subsequent tectonic inversion.
该盆地扩展发育，随后构造反转。

structural differentiation np. 构造分异

例：The differential salt tectonic deformation is mainly represented by structural differentiation at different levels, and structural zonation and segmentation in transverse and longitudinal sections.

盐层构造变形的差异主要表现为不同层次的构造分异，以及横断面和纵断面的结构分带性和分段性。

tectonic deformation *np.* 构造变形

例：Tectonic deformation is complicated and the main tectonic style is compressional reversing fracture.

这种构造变形十分复杂，主要构造形式为压性逆冲断裂。

periods of tectonic activity *np.* 构造活动期

例：Deltas, foreset bedding, reefs, sandbars, periods of non-depositon or quiet deposition, periods of tectonic activity, etc., may be identified by the aggregate of subtle evidences.

三角洲、前积层、礁体、沙洲、非沉积期或平静沉积期、构造活动期等可通过各种微妙迹象综合识别。

neotectonic movement *np.* 新构造运动

例：Tectonic landform is a dynamic geomorphological type, which is directly controlled by neotectonic movement.

构造地貌是指由新构造运动直接形成的一种动态的、积极活跃的地貌类型。

inversion structure *np.* 反转构造

例：Structural characteristics of inversion structure, fold association style, thrust fault zone upright strata developed belt shown in great profiles were also discussed.

对剖面中反映的反转构造、褶皱组合样式、逆冲断裂带、垂直地层发育带的构造特征进行了探讨。

diapir structure *np.* 底辟构造

例：The neotectonic movement not only controlled the formation of diapir structure belts and their echelon arrangement, but also dominated the petroleum accumulation and distribution.

新构造运动不仅影响了底辟构造带发育及其雁行排列特征，同时还影响了盆地油气的聚集和分布。

homogenous *adj.* 均质的

例：If the outer portion of the earth's crust were <u>homogenous</u>, this procedure would provide us with no information which we did not already have.

如果地壳外部是均质的，这个方法就不会给我们提供任何我们还没有得到的信息。

heterogeneous *adj.* 非均质的

例：The quantitative evaluation and productivity prediction of <u>heterogeneous</u> carbonate reservoir is keeping a hard-solved problem in well logging analysis.

非均质碳酸盐岩储层的定量评价和产能预测一直是测井分析中难以解决的问题。

sedimentary facies *np.* 沉积相

例：The types of hydrocarbon pools are mainly controlled by the extension of rifted basin, contemperaneous faults and <u>sedimentary facies</u>.

油气藏的类型主要受裂陷盆地的伸展程度、同生断层及沉积相带的影响。

depositional model *np.* 沉积模式

例：Traditional coal <u>depositional model</u> can hardly explain the cause of this extra-thick coal seam.

传统的煤沉积模式难以解释超厚煤层的成因。

plastic flow *np.* 塑性流动

例：Folding in layered rocks may give rise to <u>plastic flow</u>, to microscopic fracturing and slipping, and to the sliding of one layer over another.

层状岩石中的褶皱作用可导致塑性流动，引起细微的断裂和滑动，并引起一个岩层在另一个岩层上的滑动。

expulsion mechanism *np.* 排烃机制

例：The seepages in the mine at near-atmospheric pressure may not require an <u>expulsion mechanism</u> such as functions at much greater depths in petroleum-bearing basins.

在接近大气压的情况下，矿井中的石油渗出作用可能不像含油盆地埋藏很深的层位那样需要排烃机制。

hydrocarbon generation potential *np.* 生烃潜力

例：There are quite large differences in distribution, geological and geochemical characteristics and hydrocarbon generation potential of source rocks between different types of basins.

不同类型盆地的烃源岩的分布、地质地化特征和生烃潜力差别较大。

active source rock *np.* 有效烃源岩

例：A pod of active source rock indicates that a contiguous volume of organic matter is creating petroleum.

有效烃源岩是指特定时期内其内部的有机质正在生成油气。

spent source rock *np.* 过成熟烃源岩

例：The volume or pod of active source rock is determined by mapping the organic facies and considered to be the presently active, inactive, or spent source rock using organic geochemical data displayed as geochemical logs.

有效烃源岩是根据有机相来确定的，根据地球化学测井显示的有机地球化学数据可以确定烃源岩是未成熟的、成熟的或过成熟的。

thermal maturity *np.* 热成熟度

例：The stable relationship between the degree of compaction and thermal maturity has important practical significance in the prediciton of organic maturity of a sedimentary basin ahead of drilling.

研究压实度与热成熟度之间的稳定关系，对于钻井前预测沉积盆地的有机成熟度具有重要的现实意义。

time span *np.* 时间跨度

例：For a petroleum system whose overburden rock has been deposited over a broad area , the time span over which petroleum generation—migration—accumulation occurs is quite large.

对于上覆岩石沉积面积大的含油气系统，油气生成—运移—聚集的时间跨度相当大。

relative age dating *np.* 相对年龄测定

例：Relative age dating of sedimentary rocks is often made by studying fossils, trace fossils and microfossils.

沉积岩的相对年龄测定通常通过研究化石、痕迹化石和微化石来进行。

water-storage capacity *np.* 地层储水能力

例：Porosity is important for water-storage capacity, but for water to flow through rocks, the pore spaces must be connected.

孔隙度对地层储水能力尤为重要，但要让水能够从岩石中流出，孔隙之间必须相互连通。

porosity *n.* 孔隙度

例：The proportion of empty space in a rock is known as its porosity.

岩石中孔隙空间的体积比是岩石的孔隙度。

total porosity *np.* 总孔隙度

例：Total porosity is the ratio of the total void space in the rock to the bulk volume of the rock.

总孔隙度是岩石中孔隙的总体积与岩石总体积之比。

effective porosity *np.* 有效孔隙度

例：Effective porosity is the ratio of the interconnected void space in the rock to the bulk volume of the rock, each expressed in percent.

有效孔隙度是互相连通的孔隙的体积与岩石总体积之比。

pore pressure *np.* 孔隙压力

例：Water also can infiltrate the rock and significantly increase the rock's pore pressure, which also can set off or hasten the arrival of a temblor.

水还可以渗透进入岩石，增加岩石的孔隙压力，也会引发或加速地震的到来。

pore evolution *np.* 孔隙演化

例：Diagenesis controls the pore evolution, with symbolic meaning, it is very important for the exploration of the oil and gas.

成岩作用影响孔隙演化，对油气开采具有重要的指示作用。

fracture geometry *np.* 裂缝形态

例：Filtration rate of fracturing fluid into formation is one of key factors to determine fracture geometry in the design of fracturing and the evaluation after fracturing.

进行压裂设计和压裂后评估分析时，压裂液向地层滤失的速度是确定裂缝形态的关键因素之一。

fractured porous media *np.* 裂缝型孔隙介质

例：The equivalent medium model of fractured porous media is an important method to study oil and gas exploration and development of fractured reservoirs.

裂缝型孔隙介质的等效介质模型是研究裂缝性油气藏的油气勘探与开发的重要方法。

fracture pressure *np.* 裂缝压力

例：One of the important parts of formation pressure detection is to determine formation fracture pressure, which is important for drilling, well completion and fracturing.

地层压力检测的一个重要方面是确定地层裂缝压力，这对钻井、完井和压裂等施工非常重要。

permeability *n.* 渗透率

例：Permeability is the ability of the formation to conduct fluids (formation fluid conductance capacity).

渗透率是岩层传导流体的能力（岩层导流能力）。

absolute permeability *np.* 绝对渗透率

例：Absolute permeability is the measurement of the permeability conducted when a single fluid, or phase, is present in the rock.

岩石中只存在单一相或流体时所测得的渗透率是绝对渗透率。

relative permeability *np.* 相对渗透率

例：On the contrary, oil-phase relative permeability declines quickly, while aqueous phase relative permeability goes up slowly.

反之，油相相对渗透率下降快，则水相相对渗透率上升慢。

interfacial tension *np.* 界面张力

例：As the heptane in oil phase increases, the <u>interfacial tension</u> decreases.

随着油相中庚烷含量增多，界面张力减小。

fault *v.* 断裂

例：Crestal areas may be so extensively <u>faulted</u> that lines across them may be non-definitive.

脊地可能存在大面积断裂，因此穿越这些区域的测线路可能分辨不清晰。

angular unconformity *np.* 角度不整合

例：The formation of an <u>angular unconformity</u> started with the deposition of horizontal sediment layers as ancient seas covered the earth.

角度不整合的形成始于古代海洋覆盖地球时水平沉积层的沉积。

trapping mechanism *np.* 圈闭机理

例：Deep-basin oil and gas accumulation is a kind of unconventional hydrocarbon reservoirs formed under unique conditions, with special <u>trapping mechanism</u>, distribution pattern and inverted fluid contact relationship.

深盆油气藏是一种非常规油气藏。其形成条件特殊，具有特殊圈闭机理、分布规律及流体关系倒置等特征。

Plate Tectonics *np.* 板块构造学说

例：While Continental Drift and Sea-Floor Spreading are two theories of how the earth's crust was formed, <u>Plate Tectonics</u>, a combination of the former two theories, is more widely accepted.

大陆漂移学说和海底扩张学说是有关地壳形成的两种理论，而板块构造学说结合了这两种理论，为更多人所接受。

oceanization *np.* 地壳海洋化

例：Both sea-floor spreading and <u>oceanization</u> may lead to the formation of oceanic crust.

海底扩张和地壳海洋化，均可能形成海洋地壳。

Section IV

Oil and Gas Reservoir
油气藏类型

oil reservoir np. 油藏

例：An oil reservoir generally contains three fluids-gas, oil, and water-with oil the dominant product.

油藏通常有三种流体，油、气和水，其中油是主要产品。

gas reservoir 气藏

例：In addition to its occurrence as a cap or in solution, gas may accumulate independently of the oil; if so, the reservoir is called a gas reservoir.

除了以溶解气或气顶的形式存在，天然气也可以独立存在，这种情况就是气藏。

low-permeability reservoirs np. 低渗透油藏

例：Swelling, cracking, dispersion, migration and plugging of clay minerals are the main factors of the reservoir damage in high-pressure low-permeability reservoirs.

黏土矿物的膨胀、破裂、分散、运移和堵塞是对高压低渗透油藏储层造成损害的主要因素。

tight reservoir np. 致密油藏

例：Hydrocarbon production from tight reservoirs can be difficult without stimulation operations.

如果不进行增产作业，致密油藏的油气生产将非常困难。

fault block reservoir np. 断块油藏

例：Because of fault development and size shape, the exploitation of fault block reservoir is very difficult.

断块油藏由于其断层发育和大小形状差异很大，开发难度巨大。

carbonate oil reservoirs *np.* 碳酸盐岩油藏

例：The key of searching for carbonate oil reservoirs is to find large water-eroded caves filled with liquids.

寻找充满流体的大型溶蚀溶洞是寻找碳酸盐岩油藏的关键。

bottom water reservoir *np.* 底水油藏

例：The problem of low permeability bottom water reservoir fracturing hasn't been solved well and the related reports is very few.

低渗透底水油藏压裂问题尚未得到很好解决，且相关报道很少。

gas condensate reservoir 凝析气藏

例：The liquid fraction is known as "condensate" and the type of reservoir is known as a "gas condensate reservoir".

这部分液态油即"凝析油"，这种类型的气藏就是通常所说的"凝析气藏"。

porous reservoir rock *np.* 孔隙性储油岩

例：Oil lay trapped in porous reservoir rocks, such as sandstone or limestone, far beneath the surface.

石油圈闭在远离地面以下的孔隙性储油岩中，如砂岩或石灰岩。

extra-low permeability reservoir *np.* 超低渗储层

例：Throat type in extra-low permeability sandstone reservoir is versatile and minute.

超低渗砂岩储层喉道类型多样。

saturation *n.* 饱和度

例：At a given saturation nonwetting permeabilities for an unconsolidated porous medium are higher than the drainage permeabilities.

在一定的饱和度下，疏松多孔介质的非润湿渗透率高于排水渗透率。

water saturation *np.* 含水饱和度

例：Tight gas wells are generally characterized by low porosity, low permeability and high water saturation.

致密气井普遍具有低孔、低渗和高含水饱和度的特点。

oil saturation *np.* 含油饱和度

例：The oil saturation correlates positively with the permeability.
含油饱和度与渗透率呈正相关关系。

impervious/impermeable layer of rock *np.* 不渗透的岩层

例：Instead, oil upward migration was stopped by an impervious or impermeable layer of rock.
石油在向上运移时受到一些不渗透的岩层阻挡。

lithostratigraphic reservoirs *np.* 地层岩性油气藏

例：In large scale, the flexure slope break controlled the distribution of lithostratigraphic reservoirs.
挠曲坡折在很大程度上控制了地层岩性油气藏的分布。

source bed/source rock *np.* 生油层/烃源岩

例：The oil originates in a source bed, and a marine shale is thought to be a common source rock.
石油源于生油层，海相页岩被认为一种常见的生油岩。

caprock/seal rock *np.* 盖岩

例：Two common sedimentary rocks that can be caprocks are shale and salt.
两种常见的可以生成盖层的沉积岩是页岩和盐层。

cover bed *np.* 盖层

例：Cover bed is impermeable while the reservoir rock is permeable.
盖层是不渗透的岩层而储层是渗透性岩层。

reservoir bed *np.* 储层

例：An impermeable layer must occur above a reservoir bed.
储层之上必须有不渗透层。

major reservoir rock *np.* 主要储层岩

例：Major and minor reservoir rocks are determined from the percentage of in-place petroleum that originated from a particular pod of active source rock.

主要和次要储层岩是根据烃源岩原地生成的石油百分比确定的。

minor reservoir rock *np.* 次要储集岩

例：Reservoir rocks that contain minor amounts of in-place hydrocarbons are the minor reservoir rocks.

地质储量少的碳氢化合物储集岩是次要储集岩。

overburden rock *np.* 上覆岩石

例：The function of the overburden rock is more subtle because, in addition to providing the overburden necessary to thermally mature the source rock, it can also have considerable impact on the geometry of the underlying migration path and trap.

上覆岩石的作用更为微妙，除了提供源岩热成熟所需的覆盖层，上覆岩石还能对其下运移路径和圈闭的几何结构产生相当大的影响。

trap *n.* 圈闭

例：A trap is the place where oil and gas are barred from further movement.

圈闭是阻止石油和天然气进一步运移的地方。

structural traps *np.* 构造圈闭

例：Structural traps are traps that are formed because of a deformation in the rock layer that contains the hydrocarbons.

构造圈闭是由于含有碳氢化合物的岩层变形而形成的圈闭。

stratigraphic trap *np.* 地层圈闭

例：The trapping mechanism of stratigraphic traps is from stratigraphic rather than structural causes.

地层圈闭主要是由地层因素而不是构造因素所造成的。

primary stratigraphic trap *np.* 原生地层圈闭

例：Primary stratigraphic traps are formed by deposition of the reservoir rock, eg, river channel sandstone or limestone reef.

原生地层圈闭是由储层岩石（如河道砂岩或石灰岩礁）沉积形成的。

secondary stratigraphic trap *np.* 次生地层圈闭

例：A <u>secondary stratigraphic trap</u> is formed by an angular unconformity.
次生地层圈闭由角度不整合形成。

structural-lithologic trap *np.* 构造—岩性圈闭

例：The stratigraphic trap is the main type in the upper position of the slope south of the slope break, while the structural trap and the <u>structural-lithologic trap</u> dominate in the vicinity of the slope break.
坡折带以南的斜坡上部以地层圈闭为主；坡折带附近及深洼区以构造圈闭和构造—岩性圈闭为主。

subtle trap *np.* 隐蔽圈闭

例：<u>Subtle traps</u> are possibly forecasted using seismic information.
可以利用地震信息预测隐蔽圈闭。

combination trap *np.* 复合圈闭

例：<u>Combination traps</u> have both structural and stratigraphic trapping elements such as the Hugoton-Panhandle field.
复合圈闭具有构造圈闭和地层圈闭两种要素，如休格顿—潘汉尔德域。

fossil coral reef *np.* 古珊瑚礁

例：The most obvious forms of stratigraphic trap are <u>fossil coral reefs</u>.
最明显的地层圈闭形式是古珊瑚礁。

Wedging/ pinching out *np.* 尖灭

例：The commonest form of stratigraphic trap is the <u>wedging</u> or <u>pinching out</u> of a sand.
地层圈闭最常见的形态是砂岩尖灭。

unconformity *n.* 不整合

例：The gas and oil form below the <u>unconformity</u> in a source rock such as black shale.
油气形成于黑色页岩等烃源岩的不整合面下方。

interstitial water *np.* 隙间水

例：While some interstitial water is always present in the oil zone, the latter is not always underlain by a continuous body of water.

油层中虽然总是有一定量的隙间水，但其下面并不总是存在连续水体。

connate water *np.* 原生水

例：The percentage of connate water can be measured by means of electric log surveys.

原生水的含量可以用电测方法加以测定。

pore fluids *np.* 孔隙流体

例：Knowledge of the chemistry of pore fluids and their effect on the stability of minerals can be used to predict where porosity may be destroyed by cementation, preserved in its original form, or enhanced by solution of minerals by formation waters.

了解孔隙流体的化学性质及其对矿物稳定性的影响，可以预测孔隙度会因胶结作用而破坏、保持其原始形态、或因地层水溶解矿物而有所增强的区域。

oil-dissolved phase *np.* 油溶相

例：The petroleum migration mainly in an oil phase, as for natural gas migration, it can occur in water-solution, oil-dissolved, free gas and diffusion phase.

石油运移的相态主要是油相，天然气可以是水溶相、油溶相、气相和扩散相。

water-dissolved phase *np.* 水溶相

例：Gas in water-dissolved phase, in oil-dissolved phase and in diffusive phase can change into gas in free phase.

天然气既可以由水溶相、油溶相和扩散相向游离相转变，又可以由游离相向水溶相、油溶相和扩散相转变。

salt dome *np.* 盐丘

例：A salt dome formed when a mass of salt flows upwards under the pressure resulting from the weight of the overlying sediments.

上覆沉积物所产生的压力使大量的盐向上涌，就形成了盐丘。

Part I Petroleum Geology and Exploration 石油地质勘探

fault trap *np.* 断层圈闭

例：A <u>fault trap</u> occurs when the formations on either side of the fault have been moved into a position that prevents further migration of petroleum.

当断层两侧的地层移动到阻止石油进一步运移的位置时，就会出现断层圈闭。

anticline trap *n.* 背斜圈闭

例：An <u>anticline trap</u>, the commonest form of petroleum accumulation, is an upward fold in the layers of rock, much like an arch in a building.

背斜圈闭是最常见的石油储集构造，其岩石向上折起，形状类似拱形建筑物。

gas cap *np.* 气顶

例：The gas is the lightest and goes to the top of the trap to fill the pores of the reservoir rock and form the free <u>gas cap</u>.

气体最轻，进入圈闭顶部，填充储层岩石的孔隙，形成游离气顶。

aquifer *n.* 含水层

例：Where a considerable volume of water does underlie the oil in the same sedimentary bed it is referred to as the "<u>aquifer</u>".

在同一沉积岩中，石油下面如果确实存在数量可观的水体，就称之为"含水层"。

bottom waters *np.* 底水

例：This heavy tar mat is produced by the degradation of oil as <u>bottom waters</u> move beneath the oil-water contact.

这种重油垫是由于底水在油水界面下方移动时，油的降解而产生的。

edge zone *np.* 边缘带

例：The zone immediately beneath the hydrocarbons is referred to as the bottom water, and the zone of the reservoir laterally adjacent to the trap as the <u>edge zone</u>.

直接位于碳氢化合物下方的区域称为底水，与圈闭横向相邻的储层区带称为边缘带。

reservoir space *np.* 储层空间

例：Once a trap has been filled to its spill point, further storage or retention of hydrocarbons will not occur for lack of reservoir space within that trap.

一旦圈闭充满到溢出点，圈闭内的储存空间不足，就无法再储存或保留油气。

prospect *n.* 远景圈闭

例：Whenever plays or prospects are discussed, economically producible hydrocarbons are implied or anticipated.

一旦讨论含油气区带或远景圈闭阶段，就意味着已经拥有或极有希望获得经济可采的油气。

play *n.* 含油气区带

例：Economic considerations are unimportant in sedimentary basin and petroleum system investigations, but are essential in play or prospect evaluation.

经济评价在沉积盆地和含油气系统勘探研究阶段并不重要，但在含油气区带或远景圈闭阶段至关重要。

oil-gas accumulation zone *np.* 油气聚集带

例：Relationship between sedimentary facies and hydrocarbon distribution shows that most favorable oil-gas accumulation zones occur in the mouth bar, distal bar and sand sheet.

该区沉积相与油气分布关系分析表明，油气聚集带主要分布在河口坝、远砂坝和席状砂微相。

petroleum system *n.* 油气系统

例：The term petroleum system accounts for the interdependence of the essential processes of trap formation and generation-migration-accumulation of petroleum.

油气系统一词解释了圈闭形成和油气生成—运移—聚集的基本过程之间的相互依存关系。

Section V

Fundamental Exploration Methods
基本勘探方法

subsurface resource 地下资源

例: In the exploration for <u>subsurface resources</u> the methods are capable of detecting local features of potential interest that could not be discovered by any realistic drilling programme.

在地下资源勘探中，用这些方法能发现实际钻井都无法发现的潜在价值的局部特征。

oil and gas exploration *np.* 油气勘探

例: Low permeability reservoir is one of the main <u>oil and gas exploration</u> areas in China.

低渗透储层是我国目前油气勘探的主要目标之一。

favorable exploration area *np.* 有利勘探区

例: The uplift zone, slop zone and inherited rift are <u>favorable exploration area</u> for progressive exploration and development.

隆起带、斜坡带和继承性裂谷是推动油气勘探开发的有利勘探区。

trail–building or trail–cutting crew *np.* 修路或开路小组

例: In areas of difficult terrain or heavy vegetation, <u>trail-building or trail-cutting crews</u> may be required.

在地表复杂或植被繁茂的地区可能需要修路或开路小组。

amphibious vehicle *np.* 水陆两用车

例: In swampy areas the drills are often mounted on <u>amphibious vehicles</u>.

在沼泽地区，钻机通常安装在水陆两用车上。

geological exploration *np.* 地质勘探

例：Shale (mudstone) densimeter is a special instrument used to measure shale (mudstone) density during drilling in <u>geological exploration</u> for oil and gas.

页（泥）岩密度计是油气地质勘探钻井作业过程中测量页(泥)岩密度的专用计量器具。

geochemical exploration *np.* 地球化学勘探

例：Oil and gas <u>geochemical exploration</u> is an effective method to search for CO_2 gas reservoirs.

油气地球化学勘探是寻找 CO_2 气藏的有效方法。

geophysical exploration *np.* 地球物理勘探

例：The seismic method of <u>geophysical exploration</u> is the most frequently used method in petroleum exploration.

地球物理勘探的地震法是石油勘探最常用的方法。

microbial prospecting *np.* 微生物勘探

例：<u>Microbial prospecting</u> of oil and gas (MPOG) is of many advantages such as direct display, effectiveness, and less multi-interpretation.

油气微生物勘探（MPOG）具有显示直接、效率高以及重复解释少等诸多优点。

reconnaissance survey *np.* 普查勘探

例：<u>Reconnaissance surveys</u> are often carried out in the air because of the high speed of operation.

地质普查作业速度快，通常在空中进行。

airborne survey *np.* 航空勘探

例：The speed of operation of <u>airborne surveys</u> makes the magnetic surveys very attractive in the search for types of ore deposit that contain magnetic minerals.

航空勘探的高速度使磁法勘探在寻找含有磁矿物矿床方面很有吸引力。

marine surveying *np.* 海洋勘探

例：<u>Marine surveying</u> is obviously slower than aeromagnetic surveying.

海洋勘探明显比航磁勘探慢。

offshore search for oil and gas *np.* 近海油气勘探

例：In the <u>offshore search for oil and gas</u> an initial gravity reconnaissance survey may reveal the presence of a large sedimentary basin that is subsequently explored using seismic methods.

近海寻找油气最初使用重力普查勘探方法，可以发现大型沉积盆地，然后再用地震方法进一步勘探这个盆地。

deep-water exploration *np.* 深水勘探

例：<u>Deep-water exploration</u> and production have dramatically increased over the past few years.

最近几年，深水勘探和生产势头增长迅速。

natural field method *np.* 自然场方法

例：The <u>natural field methods</u> utilize the gravitational, magnetic, electrical and electromagnetic fields of the Earth, searching for local perturbations in these naturally occurring fields.

自然场方法利用地球的重力场、磁力场、电场和电磁场，调查这些自然发生场的局部波动。

artificial source method *np.* 人工源方法

例：<u>Artificial source methods</u> involve the generation of local electrical or electromagnetic fields that may be used analogously to natural fields, or the generation of seismic waves to provide information on the distribution of geological boundaries at depth.

人工源方法是生成局部电场或电磁场来模拟自然场，或者生成地震波，以提供地下地质界面的分布信息。

remote sensing technique *np.* 遥感技术

例：<u>Remote sensing technique</u> helps record the terrains, rock distributions, geological images and tectonic phenomena, etc.

遥感技术能够将各种地形、岩石分布、地质形象和构造现象等记录下来。

surveyor *n.* 勘测员

例：The surveyor should indicate in his data and maps the locations of all important features such as streams, buildings, roads, fences, etc.

勘测员应在其资料和图纸上标明所有诸如河流、建筑物、道路、围墙等的位置。

planetable alidade *np.* 平板照准仪

例：Depending on the sort of terrain being traversed, planetable alidade are sometimes used.

根据所穿越地形的类型，有时会使用平板照准仪。

gravity *n.* 密度

例：In oil reservoirs, these fluids occur in different phases because of the variance in their gravities.

油藏中这三种流体由于密度不同而呈现不同的相态。

gravity survey *np.* 重力勘探

例：Seismic surveys are often complemented with data from magnetic and gravity surveys and actual physical data garnered from exploratory wells.

地震勘探通常参照磁力和重力勘探的数据以及从探井中获得的实际物理数据。

gravitational field *np.* 重力场

例：Any lateral density variation associated with a change of subsurface geology results in a local deviation in the gravitational field.

地下地质发生变化，横向密度就会发生变化，导致重力场出现局部偏差。

gravimeter *n.* 重力仪

例：It is a reasonably simple matter to prospect for salt domes and their accompanying petroleum deposits by means of a gravimeter.

用重力仪勘探盐丘及其伴生石油矿很简单。

gravitational intensity *np.* 重力强度

例：Variations in gravitational intensity can be attributed to a number of causes.

重力强度的变化有许多原因。

latitude effect *np.* 纬度效应

例：One major cause of change in gravitational intensity is entitled the latitude effect.

重力强度变化的一个主要原因是纬度效应。

gross gravity separation *np.* 总重力分离

例：Not only does a gross gravity separation of gas and oil occur within a reservoir but more subtle variation may also exist.

储层内不仅会发生天然气和石油总重力分离，还可能会有更微妙的变化。

microgravimetric technique *np.* 微重力测量技术

例：Microgravimetric techniques find occasional geotechnical applications in the location of subsurface voids.

微重力测量技术有时也用于地质上寻找地下孔隙。

surface gravity survey *np.* 地面重力测量

例：It is probable that specialized borehole gravimeters and surface gravity surveys will be increasingly used in the analysis of the interwell regions of hydrocarbon reservoirs.

对油气藏井间区域进行分析，可能会更多地使用专业井筒重力仪和地面重力测量。

portable instrument *np.* 便携式仪器

例：Another important recent development in gravity surveying is the design of a portable instrument capable of measuring absolute gravity with high precision.

最近在重力测量领域，另一个重要进展是设计了一种便携式仪器，能够高精度测量绝对重力。

geophysical anomaly *np.* 地球物理异常

例：The variation attributed to a localized subsurface zone of distinctive geophysical property and possible geological importance is known as a geophysical anomaly.

地球物理异常指局部地下地层出现偏差，该地层具有明显的地球物理特征或地质上可能发生重大变化。

positive and negative gravity anomaly 正重力异常和负重力异常

例：Very early marine surveys indicated the existence of large positive and negative gravity anomalies associated with island arks and oceanic trenches respectively.

早期海洋勘探表明，大型正重力异常和负重力异常与岛屿和海沟相关。

magnetic exploration *np.* 磁法勘探

例：Magnetic exploration is a geophysical method on magnetic change observed in the underground medium.

磁法勘探是一种地球物理方法，探测地下介质中磁力的变化。

magnetic susceptibility *np.* 磁化率

例：The magnetic method is very suitable for locating buried magnetite ore bodies because of their high magnetic susceptibility.

因磁铁矿磁化率高，磁法勘探非常适合寻找磁铁矿。

magnetic field *np.* 磁场

例：The salt property of negative magnetic susceptibility causes a local decrease in the strength of the Earth's magnetic field in the vicinity of a salt dome.

盐具有负磁化率特点，使盐丘附近的地球磁场强度减弱。

fluxgate magnetometer *np.* 饱和式磁力仪

例：The magnetotelluric field is measured by its inductive effect on a coil about a metre in diameter or by use of a sensitive fluxgate magnetometer.

通过直径约一米的线圈感应效应或高灵敏度饱和式磁力仪，可以测量大地电磁场。

proton magnetometer *np.* 质子磁力仪

例：Base station readings are not necessary for monitoring instrumental drift as fluxgate and proton magnetometers do not drift but may be used to monitor diurnal variation.

基点读数对于监测仪器漂移不是必需的，因为饱和式磁力仪和质子磁力仪不会漂移，但基点读数可以用来监测日变化。

aeromagnetic survey *np.* 航空磁测

例：The initial search for metalliferous mineral deposits often utilizes aeromagnetic and electromagnetic surveying.

初期探测金属矿床一般用航空磁测和电磁勘探。

terrain photograph *np.* 地形摄影

例：Terrain photographs are taken simultaneously so that the location can subsequently be determined by reference to topographic maps.

地形拍摄同时进行，以便随后根据地形图确定位置。

linear magnetic anomaly *np.* 线性磁异常

例：A significant contribution in this field was the discovery of linear magnetic anomalies within the oceanbasins, which led directly to the theory of sea floor spreading.

该领域重大贡献是发现了海洋盆地线性磁异常，直接导致了海底伸展理论形成。

electromagnetic field *np.* 电磁场

例：The induction of current flow results from the magnetic component of the electromagnetic field.

电流感应是由电磁场的磁分量造成的。

magnetotelluric field *np.* 大地电磁场

例：Prospecting using magnetotelluric fields is more complex than the telluric method as both the electric and magnetic fields must be measured.

大地电磁场勘探比大地电流法勘探复杂得多，因为电流和电磁场都必须测量。

magneto telluric survey *np.* 大地电磁勘探

例：Interpretation of magneto telluric survey data is most reliable in the case of horizontal layering.

在水平分层的情况下，大地电磁勘探数据的解释是最可靠的。

primary electromagnetic field *np.* 一次电磁场

例：Primary electromagnetic fields may be generated by passing alternating current through a small coil made up of many turns of wire or through a large loop of wire.

交流电通过大线圈或通过多匝导线组成的小线圈时，就会产生一次电磁场。

secondary electromagnetic field *np.* 二次电磁场

例：The response of the ground is the generation of secondary electromagnetic fields.

地面响应就会产生二次电磁场。

electromagnetic method 电磁法勘探

例：Exploration for ore bodies is mainly carried out using electromagnetic surveying methods.

矿体勘探主要采用电磁测量方法。

airborne EM method *np.* 航空电磁勘探法

例：Airborne EM methods are widely used in prospecting for conductive ore bodies.

航空电磁勘探法已广泛应用于导电矿体勘探。

electrical method 电法勘探

例：Electrical methods are suitable for the location of a buried water table because saturated rock may be distinguished from dry rock by its higher electrical conductivity.

电法勘探适合寻找潜水面，因为饱和岩层导电率高，可以和干岩层区分开来。

telluric current *np.* 大地电流

例：Much greater penetration can he achieved by making use of the natural Earth currents (telluric currents) generated by the motions of charged particles in the ionosphere.

自然大地电流（大地电流）因电离层中带电粒子的运动而产生，可以穿透更深的地层。

artificially-generated *adj.* 人工产生的

例：In the resistivity method, <u>artificially-generated</u> electric currents are introduced into the ground and the resulting potential differences are measured at the surface.

用电阻率法可将人工产生的电流输入地下，并在地表测出电位差。

conductivity *n.* 导电性

例：Similarly seismic or electrical methods are suitable for the location of a buried water table because saturated rock may be distinguished from dry rock by its lower seismic velocity and higher electrical <u>conductivity</u>.

同样，地震或电法勘探也适于测定地下水位，因为饱和岩石可能因其较低的地震速度和较高的导电性而有别于干岩石。

electrical resistivity *np.* 电阻率

例：A salt dome, however, possesses an anomalously high <u>electrical resistivity</u> and electric currents preferentially flow around and over the top of such a structure rather than through it.

然而，盐穹电阻率很高，电流会在其周围流动，或流经其顶部，而非在其中穿行。

resistivity meter *np.* 电阻率仪

例：Most modern <u>resistivity meters</u> employ low frequency alternating current rather than direct current for two main reasons.

现代大多数电阻率仪采用低频交流电而不是直流电，主要有两个原因。

self-potential method *np.* 自然电位法

例：The <u>self-potential method</u> makes use of natural currents flowing in the ground that are generated by electrochemical processes to locate shallow bodies of anomalous conductivity.

自然电流由电化学作用产生，自然电位法利用地下自然电流定位电导率异常的浅层物体。

potential difference *np.* 电位差

例：Shallow features are normally investigated using artificial field methods in which an electrical current is introduced into the ground and <u>potential differences</u>

between points on the surface are measured to reveal anomalous material in the subsurface.

浅地表勘探通常使用人工场方法，将电流输入地下，并测量地表各点之间的电位差，以发现地下异常物质。

mean potential gradient *np.* 平均电位梯度

例：Although variable in both their direction and intensity, telluric currents cause a <u>mean potential gradient</u> at the Earth's surface.

虽然在方向和强度上有变化，但大地电流在地球表面生成平均电位梯度。

dielectric property *np.* 介电性

例：Penetration depth was a function of <u>dielectric property</u> positively correlated with temperature.

穿透深度是介电性的函数，与温度正相关。

induced polarization *np.* 激发极化

例：<u>Induced polarization</u> is observed when a steady current through two electrodes in the Earth is shut off: the voltage does not return to zero instantaneously, but rather decays slowly, indicating that charge has been stored in the rocks.

切断通过地球两个电极的稳定电流，可以观察到激发极化现象：电压不会瞬间恢复到零，而是缓慢衰减，表明电荷已经储存在岩石中。

electronic positioning system *np.* 电子定位系统

例：Where available, <u>electronic positioning systems</u> are employed.

可能的情况下就采用电子定位系统。

anomalous material *np.* 异常物质

例：These currents extend to great depths within the Earth and, in the absence of any electrically <u>anomalous material</u>, flow parallel to the surface.

这些电流流向地球深处，在没有遇到异常物质的情况下，平行于地表流动。

fence diagram *np.* 栅状图

例：A <u>fence diagram</u> is used to show how wells correlate in three dimensions.

栅状图用于显示（油）井之间油层的三维连通状况。

exploratory well 探井

例：Investigation of sedimentary basins requires a low-density information grid that covers a large area, such as widely spaced seismic lines, a few strategically placed <u>exploratory wells</u>, and small-scale geologic maps.

沉积盆地勘探阶段需要获取网格密度低、覆盖范围大的资料，例如采用大间距地震测线、几个战略部署的探井和小比例尺地质图。

radius of investigation *np.* 探测/勘探半径

例：The <u>radius of investigation</u> depends on the extent and scope of formation damage.

勘探半径取决于地层损害的程度和范围。

sonic log/acoustic log *np.* 声波测井

例：<u>Sonic logs</u> are often not run in the upper part of a borehole and consequently many of the most important effects cannot be modeled correctly.

声波测井通常不在井筒上段进行，因此许多效应无法正确模拟。

geophysical well log *np.* 地球物理测井

例：Detailed knowledge of the mineralogy of reservoirs is essential for the accurate interpretation of <u>geophysical well logs</u> through reservoirs.

储层矿物学知识细节对于准确解释储层地球物理测井至关重要。

continuous velocity log (CVL) *np.* 连续速度测井

例：When a well has been drilled a <u>continuous velocity log (CVL)</u> may be run in the hole.

钻完井之后，即可在井中进行连续速度测井（CVL）。

geophysical wireline well log *np.* 地球物理电缆测井

例：Nor could any finds be evaluated effectively without <u>geophysical wireline well logs</u> to measure the lithology, porosity, and petroleum content of a reservoir.

如果没有地球物理电缆测井来测量储层的岩性、孔隙度和石油含量，也无法有效评估任何发现的圈闭。

well log/borehole log *np.* 测井曲线

例：The <u>well log</u> is recorded by an instrument which measures and integrates travel times over short portions of the wells.

用仪器将井内短距离震波传播时间测量出来并加以综合计算，记录测井曲线。

coring *n.* 取心

例：For direct quantitative measurement of porosity, reliance must be placed on formation samples obtained by <u>coring</u>.

直接定量测定孔隙度，通过取心获得地层岩样来测定才可靠。

core analysis *np.* 岩心分析

例：The invention is applicable to nuclear magnetic resonance for well logging and <u>core analysis</u> in exploration and development for oil field.

本发明适用于油田勘探开发中测井和岩心分析的核磁共振操作。

cross section *np.* 横剖面

例：<u>Cross sections</u> are useful for displaying the types and orientations of subsurface structures and formations.

横剖面对于显示地下结构和地层的类型和方向非常有用。

record section *np.* 记录剖面

例：The end result of the data processing is usually <u>record sections</u> from which an interpretation is made.

数据处理的最终结果通常是记录剖面，再据其进行解释。

geologic structure section *np.* 地质构造剖面

例：A <u>geologic structure section</u> is a diagram of the rocks and structures that would occur along one of its walls.

地质构造剖面是沿着其中一个侧壁所出现的岩石和构造的图形。

geologic map *np.* 地质图

例：A <u>geologic map</u> shows the distribution of rocks of different types and ages at the surface.

地质图显示了地表不同类型和年龄的岩石分布。

structural map *np.* 构造图

例：Four kinds of ridge like structures, saddle structural lithological ridge, fault lithological ridge, anticlinal lithological ridge and lithological ridge, can be identified from the structural map.

从该构造图上可识别出油气向西部斜坡带运移的四种脊状构造，即鞍状构造—岩性脊、断层—岩性脊、背斜—岩性脊和纯岩性脊。

Section VI

Seismic Exploration
地震勘探

seismology n. 地震学

例：<u>Seismology</u> is the branch of geology which deals with the movement of waves through the earth.

地震学是地质学的分支，研究穿越地球波的运动。

seismologist n. 地震学家

例：A complete interpretation of complex orientations of these underground layers is very difficult and occupies a large part of the time of a large number of <u>seismologists</u>.

对这些地下岩层的复杂方位进行完整解释很困难，占用了大量地震学家的时间。

seismic exploration np. 地震勘探

例：The primary purpose of <u>seismic exploration</u> is to determine the structure of the subsurface rocks.

地震勘探的主要目的是确定地下岩石结构。

marine seismic survey np. 海洋地震勘探

例：In the case of <u>marine seismic surveys</u>, fieldwork is often completed several weeks before the results can be processed and interpreted.

在海洋地震勘探中，野外作业常常完成了几星期之后，有关资料才能处理和解释出来。

seismic facies np. 地震相

例：The objective of <u>seismic facies</u> analysis is regional stratigraphic interpretation, specifically, defining the depositional environment, the lithology, and geologic history.

地震相分析的目的是区域地层解释，特别是确定沉积环境、岩性和地质历史。

seismic sequence *np.* 地震层序

例：We use seismic facies distinctions as well as unconformities to separate <u>seismic sequence</u> units.

我们利用地震相和不整合面的各种特征来划分地震层序。

seismic crew *np.* 地震队

例：Depending upon the number and depth of holes required and the ease of drilling, a <u>seismic crew</u> will generally have from one to four drilling crews.

根据所需钻井的数量和深度以及钻井的难易程度，地震队通常有一到四支钻井队。

3-D seismic survey *np.* 三维地震勘探

例：These advances combined with improvements in computational power over the past two decades have made the acquisition and processing of <u>3-D seismic surveys</u> even more feasible and economical.

这些进展加上过去二十年来计算能力的提高，使得三维地震勘探数据的采集和处理更加可行和经济。

three-dimensional coverage *np.* 三维覆盖

例：Refraction data recorded in a broadside geometry provide good <u>three-dimensional coverage</u> of western Hecate Strait.

宽边几何图形中的折射数据提供了清晰的赫卡特海峡西部的三维覆盖。

time-lapse 3-D seismic technology *np.* 时差三维地震技术

例：Two evolving technologies that can minimize risk and maximize return for deep-water investments are seafloor seismic and <u>time-lapse 3-D (or 4-D) seismic technologies</u>.

海底地震和时差三维（或四维）地震技术这两种不断发展的技术，可以将深海投资风险降到最低，并获得最大回报。

layered sequence *np.* 分层层序

例：Seismic methods are particularly well suited to the investigation of the layered sequences in sedimentary basins that are the primary targets for oil or gas.

地震勘探法特别适合研究油气基本目的层——沉积盆地中的分层层序。

seismograph *n.* 地震仪

例：The study of the form and occurrence of earthquake waves recorded by seismographs, has been the principal source of knowledge of the constitution of the interior of the earth.

对地震仪记录的地震波及其波形进行研究，已成为了解地球内部结构的主要方法。

geophone *n.* 检波器

例：Seismic surveys are divided into two categories depending on the path taken by the waves in the sediments between the explosion and the geophones.

根据波从爆炸点到检波器间的岩层中的传播路径，可以将地震勘探分为两大类。

multiple geophone *np.* 检波器组合

例：We use multiple geophones much more than multiple shots because the cost is less.

由于成本较低，我们使用检波器组合远远多于炮点组合。

geophone station *np.* 检波器站

例：The resulting seismic record is simply the superposition of those waves which are ultimately reflected back to the geophone station.

地震记录就是最终反射回检波器站的叠加波。

transmitter *n.* 发射器

例：Both transmitter and receiver can be mounted in aircraft or towed behind them.

发射器和接收器都可以安装在飞机上或挂在飞机后面。

receiver *n.* 接收器

例：The eddy currents generate their own secondary electromagnetic field which

Part I Petroleum Geology and Exploration　石油地质勘探

travels to the <u>receiver</u>.

涡流产生二次电磁场，并传播到接收器。

amplifier *n.* 放大器

例： The electrical signals from the geophone groups go to an equal number of <u>amplifiers</u>.

检波器的电子信号传输至相应的放大器。

fan shooting *np.* 扇形排列法地震勘探

例： By means of this "<u>fan-shooting</u>" it is possible to delineate sections of ground which are associated with anomalously short travel times and which may therefore be underlain by a salt body.

"扇形排列爆破"可描绘出震波穿行时间很短的地方，提示下方可能有盐体。

seismic source *np.* 震源

例： The conventional <u>seismic source</u> is the explosion of a charge of dynamite in a drilled shothole, although many other energy sources are now in use.

常规震源是在钻好的炮井中装炸药爆破，不过目前也使用许多其他能源作震源。

point source *np.* 点震源

例： We generally assume that dynamite in a single shothole constitutes a <u>point source</u>.

我们通常认为，单井中炸药构成的是一个点震源。

source truck *np.* 震源车

例： It is not uncommon to use three or four <u>source trucks</u> and combine thirty or so component sub-shots.

用三台或四台震源车，再进行大约三十次左右的重复放炮并不罕见。

vibroseis *n.* 可控震源

例： About 70% of the seismic exploration run on land today is done by <u>vibroseis</u>, a technique developed by Continental Oil.

目前陆地上大约 70% 的地震勘探由可控震源完成，这技种术是由大陆石油公司开发的。

recording truck *np.* 记录车

例： The surveyor also plans access routes so that drills, recording trucks, etc., can get to their required locations most expeditiously.

勘测员还规划了进入测线的路径，以便钻机、记录车等能够最快到达所需位置。

shot *n.* 爆炸

例： It is necessary to record both the time of the shot and the time of arrival of the waves at the geophones.

有必要同时记录下爆炸时间和波到达各个检波器的时间。

shotpoint *n.* 炮点

例： Shotpoints are usually spaced at equal intervals of about 1/4 mile (400 meters).

炮点的间距通常为 1/4 英里（400 米）。

multiple shots *np.* 组合炮点

例： With some surface sources two to four source units may be used at the same time to provide the effect of multiple shots.

某些地表震源有可能使用两到四台震源车来获得组合炮点的效应。

common-depth-point *np.* 公共深度点

例： With common-depth-point data which have a high degree of redundancy (i.e., which sample the same depth point many times) and velocity-analysis programs, interval velocities can be calculated.

根据高重复度公共深度点数据（即多次对同一深度点采样）和速度分析程序，可以计算出层速度。

grid *n.* 测网

例： In a given area the seismic lines are usually laid out in a grid.

在特定地区，地震测线通常是布置成测网的。

seismic line *np.* 地震测线

例： Often seismic lines pass close enough to wells to permit correlating the seismic horizons with geological horizons in the wells.

地震测线通常距离油井很近，以便对比地震层位与油井中的地质层位。

spread *n.* 排列

例：By spread we mean the relative locations of the shotpoint and the centres of the geophone groups used to record the energy from the shot.

我们用排列表示炮点与检波器组中心点的相对位置，检波器用于记录从炮点传来的能量。

end-on spread/end shooting *np.* 端点放炮排列

例：Occasionally the shothole is located at the end of the line of active geophone groups to produce an end-on spread.

有时将炮点置于有效检波器组排列的末端构成端点放炮排列。

in-line offset spread *np.* 离开排列

例：Sometimes in areas of exceptionally heavy ground roll the shotpoint is removed an appreciable distance along the line from the nearest active geophone group to produce an in-line offset spread.

有时，在地滚波异常严重的区域，炮点沿测线离开有效检波器组相当长一段距离，构成离开排列。

array *n.* 组合

例：The term array refers either to the pattern of a group of geophones which feed a single channel or to a distribution of shotholes or surface energy sources which are fired simultaneously.

"组合"一词是指向单个地震道输入信号的一组检波器排列，或者是指同时放炮的一些炮点或地表震源的分布。

noisy trace *np.* 噪声道

例：Placing the shotpoint close to a geophone group often results in a noisy trace.

将炮点放置在靠近检波器组的位置通常会产生噪声道。

destructive interference *np.* 相消干扰

例：A wave traveling horizontally will affect the various geophones at different times so that there will be a certain degree of destructive interference.

平行波会在不同时间段影响不同位置的检波器，因此会有一定程度的相消干扰。

sine-wave input *np.* 正弦波输入

例：The response is usually given for the case of sine-wave input as a function of frequency, apparent velocity, dip moveout or angle of approach.

响应取决于正弦波的输入，正弦波输入是频率、视速度、倾角时差或接近角的函数。

response diagram *np.* 响应图

例：Response diagrams apply equally to arrays of geophones and arrays of shotpoints (or source units).

响应图同样适用于检波器组合和炮点（震源车）组合。

dynamite *n.* 炸药

例：Usually the drilling crew places the dynamite in the holes before leaving the site.

通常，钻井组在离开现场之前要把炸药下入井里。

shooting crew *np.* 爆炸组

例：The shooting crew are responsible for loading the shotholes and setting off the dynamite.

爆炸组负责下炸药和引爆炸药。

jug hustler *np.* 放线员

例：The jug hustlers who lay out the cables place the geophones in their proper locations and connect them into the cables.

负责放大线的放线组将检波器埋置在适当位置并将其连接到大线上。

shooter *n.* 爆破员

例：He gives the signal to the shooter via telephone or radio to fire the charge.

他通过电话或无线电向爆破员发出引爆炸药的信号。

blaster *n.* 引爆机

例：On receipt of the firing signal the shooter closes two firing switches on the blaster.

收到点火信号后，爆破员关闭引爆机上的两个点火开关。

shothole n. 炮井

例：Shotholes may be drilled for the entire line before the recording even begins so that one need never wait on the drills.

甚至在记录开始之前，整个测线的炮井就已经钻完，就不需要再等待钻机了。

roll-along switch np. 多次覆盖开关

例：Following the shot, the shooters move on to the next shothole and the observer adjusts the roll-along switch so that the next geophones are connected.

放炮后，爆破员移到下一个炮孔，观察员调整多次覆盖开关，连接下一组检波器。

the maximum offset np. 最大炮检距

例：The maximum offset is the distance from energy source to most distant live geophone station.

最大炮检距是从震源到最远的有效检波器点的距离。

the minimum offset np. 最小炮检距

例：The minimum offset is the distance from energy source to the nearest live geophone station.

最小炮检距是从震源到最近的有效检波器点的距离。

split spread np. 中间放炮排列

例：A split spread does not necessarily have to be symmetrical.

中间放炮排列不一定必须对称。

trace density np. 道密度

例：This gives some degree of control of maximum offset for a given trace density (number of stations per unit distance).

在道密度（单位距离上的检波器点数）确定的情况下，可以对最大炮检距做一定程度的控制。

ground roll *np.* 地滚波

例：Ground roll should not be a determinant of minimum offset.
地滚波不应成为设计最小炮检距的决定因素。

attenuating effect *np.* 衰减作用

例：Hole depth can be considered as having an attenuating effect on ground roll.
可以认为井深对地滚有衰减作用。

high-cut frequency filter *np.* 高截频率滤波器

例：Hole depth therefore acts as a high-cut frequency filter.
井深起了高截频率滤波器的作用。

charge depth *np.* 药深

例：Sudden or anomalous changes in up-hole time can be taken as suggestive of incorrect charge depth.
井口时间的突然或异常变化可以当作药深不正确的迹象。

charge size *np.* 装药量

例：At the start of a dynamite operation recordings should be made using different charge sizes.
炸药作业开始时，应使用不同的装药量进行记录。

signal-to-noise ratio *np.* 信噪比

例：Increased charge size is not a solution to the problem of signal-to-noise ratio.
药量增加并不是解决信噪比问题的办法。

misfire *n.* 瞎炮

例：Holes where misfires occur are not reloaded and reshot.
瞎炮不会重新下炸药再放。

seismic ray *np.* 地震射线

例：Seismic rays normally propagate through salt at a higher velocity than through the surrounding sediments.
地震射线在盐层中的传播速度一般比在周围沉积中的传播速度要高。

seismic wave *np.* 地震波

例：By knowing the depth of the explosion and the velocities of <u>seismic waves</u> in different rock types, it is a comparatively simple matter to calculate depth to layers with different velocities of seismic waves.

知道了爆破的深度和不同岩石类型传播地震波的速度，再来计算具有不同地震波速岩层的深度，那就简单了。

seismic wavelet *np.* 地震子波

例：The larger the cavity the lower the frequency content of the <u>seismic wavelet</u>.

洞穴越大，地震子波的频率成分越低。

propagate *v.* 穿越

例：Seismic rays normally <u>propagate</u> through salt at a higher velocity than through the surrounding sediments.

地震射线通常以高于周围沉积物的速度穿越盐层。

traveltime *n.* 走时

例：The interval between the shot instant and the arrival of the energy at a geophone group is known as the <u>traveltime</u>.

放炮瞬间和能量到达检波器组之间的间隔时间称为走时。

layering *n.* 分层

例：If the <u>layering</u> is inclined in some way at an angle to the earth's surface, then this fact should be fairly easily detected by placing listening devices on both sides of the shot hole.

如果分层与地球表面成一定倾角，那么在炮孔两侧放置监听设备就会很容易检测到。

reflection seismology *np.* 反射地震学

例：<u>Reflection seismology</u> involves the generation of artificial seismic waves and subsequently capturing the subsurface reflections of these waves to create multi-dimensional representations of the underground geology.

反射地震学即人工生成地震波，并跟踪反射震波，以多维视角创建地下地质面貌。

reflection path *np.* 反射路径

例：The conversion of this travel time into a depth requires knowledge of the velocity with which the pulse traveled along the reflection path.

将传播时间转换为深度需要了解脉冲沿反射路径传播的速度。

reflection event *np.* 反射波

例：The location and altitude of the interface which gave rise to each reflection event are then calculated from the arrival times.

然后根据到达时间计算产生每个反射波的界面位置和高度。

reflection amplitude *np.* 反射波振幅

例：Increasing charge size increases the amplitude of source-generated noise at the same time as it increases reflection amplitude.

增加药量会在增强反射波振幅的同时增强震源产生的噪声振幅。

reflection configuration *np.* 反射结构

例：The most common types of reflection configurations consist of parallel, subparallel, or divergent reflections.

最常见的反射结构类型有平行反射、亚平行反射和发散反射。

divergent reflection *np.* 发散反射

例：Divergent reflections gradually spread out, almost always in the down-dip direction, indicating gradual basin subsidence during deposition, with gradual basinward tilting.

发散反射几乎总是在向下倾斜的方向逐渐散开，表明沉积过程中盆地逐渐沉降并逐渐向盆地内倾斜。

reflecting layer/horizon *np.* 反射层

例：An important use of synthetic seismograms is in studying the effect of changes in the reflecting layers on the seismic record.

合成地震图的一个重要用途是研究反射层变化对地震记录的影响。

relative amplitude *np.* 相对振幅

例：The amplitudes of reflections, especially relative amplitudes, provide information

for stratigraphic interpretation.
反射振幅，特别是相对振幅，为地层解释提供了信息。

primary reflection *np.* 一次反射

例：Comparison of the actual and synthetic seismograms may also help to determine which events represent primary reflections and which multiples.
将实际地震记录和合成地震记录进行比较还有助于确定哪些同相轴是一次反射，哪些是多次波。

multiple *n.* 多次波

例：Often multiples, especially short-path multiples, are ignored.
通常忽略的是多次波，尤其是短程反射多次波。

seismic refraction *np.* 地震折射

例：Seismic refraction method is a seismic acquisition method in which the incident and reflected angles are critical.
地震折射法是一种地震采集方法，其中入射角和反射角是关键。

refraction velocity *np.* 折射速度

例：Refraction velocities (if available) may help identify certain horizons.
折射速度（如果可获得）可能有助于确定某些层位。

reflection and transmission coefficient *np.* 反射和透射系数

例：Density variations are frequently ignored so that the reflection and transmission coefficients are based on velocity changes only.
密度变化常被忽略，因此反射和透射系数仅基于速度变化。

direct problem *np.* 正演问题

例：Direct problems are theoretically capable of unambiguous solution.
正演问题理论上能够得到明确的解决。

inverse problem *np.* 反演问题

例：Inverse problems suffer from an inherent ambiguity, or non-uniqueness, in the conclusions that can be drawn.

反演问题在所得结论中具有固有的模糊性或不一致性。

time-contour/isochron map *np.* 等时线/等时图

例：The time-contour, or isochron maps show all the structural features, but depth-contour maps are more convenient for exploration purposes.

虽然等时线或等时图反映了所有的构造特点，但是就勘探而言，使用起来却不如其深线图那样方便。

isopach map *np.* 等厚线图

例：The contouring of the maps can be done by the computer to allow the conversion of the times into depth and to enable the production of isopach maps.

可用计算机绘制出等时图，进行时深转换，并绘制出等厚线图。

depth value *np.* 深度值

例：Faults which have been identified on the records or cross-sections are drawn on the map and the depth values are then contoured.

在记录或横剖面上已确定的断层也勾画在图上，然后用等值线把各深度值连起来。

seismic section *np.* 地震剖面

例：In reflection seismology, the end product of seismic data processing is a seismic section that is analogous to a geologic cross section.

在反射地震学中，地震资料处理的最终成果是类似于地质横剖面的地震剖面。

geological interface *np.* 地质界面

例：Using the same principle, a simple seismic survey may be used to determine the depth of a buried geological interface.

基于相同原理，可以使用简单的地震勘探方法确定地下地质界面的深度。

continuous seismic profiling *np.* 连续地震剖面法

例：Continuous seismic profiling cannot be employed in the air.

连续地震剖面法不能在空中使用。

seismic profile *np.* 地震剖面图

例：In a vertical <u>seismic profile</u>, offset is the horizontal distance between the source and the wellhead or the surface projection of the receiver in the case of a deviated well.

在纵向地震剖面图中，偏移量是指震源与井口之间的水平距离，或者在偏离井的情况下，指接收器的表面投影。

field record *np.* 现场记录

例：When the <u>field records</u> have been processed the travel times to the reflecting or refracting formations are plotted on maps and contours drawn through equal time values.

现场记录处理后，将反射或折射时间绘制成时间图，并通过等时值绘制等高线。

wellbore seismic technique *np.* 井筒地震技术

例：Reservoir geophysical technique mainly include characterized data acquisition techniques, such as 3-D high-precision seismic technique, <u>wellbore seismic technique</u>, multi-wave/ multi-component seismic technique and 4-D seismic technique.

储层地球物理技术主要包括独特的数据采集技术，如三维高精度地震技术、井眼地震技术、多波多分量地震技术和四维地震技术。

bright spot *np.* 亮点技术

例：<u>Bright spots</u> have been used very successfully to locate gas reservoirs and free gas caps on saturated oil fields.

亮点技术已成功用于定位气藏和饱和油田的游离气顶。

reflection event *np.* 反射同相轴

例：In many areas the synthetic seismogram is a reasonable approximation to actual seismic records and is therefore useful in correlating <u>reflection events</u> with particular horizons.

在许多地区，合成地震记录接近实际地震记录，因此可用于反射同相轴与具体层位相对比。

synthetic seismogram *np.* 合成地震图

例：Synthetic seismograms are generated by calculating reflection coefficients from the sonic and density logs and then applying an ideal or real wavelet to the reflections to obtain the seismic "wiggle" traces.

通过计算声波测井和密度测井的反射系数，将理想的或真实的子波应用于反射而获得地震"摇晃"轨迹，即可生成合成地震图。

seismic data *np.* 地震数据

例：Synthetic seismograms are usually generated to compare with the actual seismic data and identify reflectors with layers and formations already known in the wellbore.

通常生成合成地震图，与实际地震数据进行比较，来识别井筒中已知地层和构造的反射层。

seismic resolving power *np.* 地震分辨率

例：The structure being sought may be beyond seismic resolving power.

地震分辨率无法分辨出要寻找的构造。

data acquisition *np.* 数据采集

例：Seismic exploration involves three phases of data acquisition, data processing and data interpretation.

地震勘探包括三个阶段，即数据采集、数据处理和数据解释。

data processing *np.* 数据处理

例：In reflection seismology, the end product of seismic data processing is a seismic section that is analogous to a geologic cross section.

在反射地震学中，地震数据处理的最终结果是地震剖面，类似于地质剖面。

data interpretation *np.* 数据解释

例：Standard polarity establishment and horizon identification are two essentials in meticulous seismic data interpretation.

标准地震极性建立和层位标定是精细地震数据解释的两项基础工作。

interpreter *n.* 解释员

例：The <u>interpreter</u> can use the sections to map subsurface geologic structure.

解释员可以利用地震剖面来绘制地下地质构造。

genetic interpretation *np.* 成因解释

例：Basin classification schemes evolved from descriptive geology to <u>genetic interpretations</u> with the advent of plate tectonics theory.

随着板块构造理论的出现，盆地分类方案从描述性地质演化发展到成因解释方面。

seismic imaging *np.* 地震成像

例：Better <u>seismic imaging</u> can reduce overall risk in the stages of exploration and appraisal.

较好的地震成像可以降低勘探阶段和评估阶段的整体风险。

4-D time-lapse imaging *np.* 四维延时成像

例：4-D seismology expands on 3-D by adding time as the fourth dimension, allowing observation of changes in subsurface characteristics over time. Since time is the fourth dimension, this technique is also called <u>4-D time-lapse imaging</u>.

四维地震学在三维的基础上增加了时间，以观察地下特征随时间的变化。由于第四维是时间，这种技术也称为四维延时成像。

transmission tomography *np.* 透射层析成像

例：Because of the constraints on velocity and depth provided by the different wave types, this algorithm substantially reduces the ambiguity between velocity and depth prevalent in reflection tomography, and also avoids the undetermined problem in <u>transmission tomography</u>.

由于不同波型对速度和深度的限制，该算法大大减少了反射层析成像中普遍存在的速度和深度不确定性，同时也避免了透射层析成像中的不确定性问题。

Section VII

Reservoir Evaluation
油气藏评价

appraisal *n.* 评价

例：The life of the field can be simplified into four stages: exploration, appraisal, development, and production.

油田生命周期可简化为四个阶段：勘探、评价、开发和生产。

geological evaluation *np.* 地质评价

例：Economic evaluation is after geological evaluation and feasibility study usually, before the next production or exploration.

经济评价通常是在地质评价和可行性研究之后，以及生产或勘探之前进行的。

productivity evaluation *np.* 产能评价

例：Gas well productivity evaluation data are applied to optimizing horizontal well design.

气井产能评价数据用于水平井优化设计。

resource assessment *np.* 资源评价

例：A method of evaluating dead oil content and estimated value available for reference in resource assessment is discussed.

探讨了一种估算死油含量和可采量的方法，可供资源评价参考。

hydrocarbon evaluation *np.* 油气评价

例：Hydrocarbon evaluation analysis is based on different level research, such as basin analysis, sedimentary sequence analysis, and petroleum system analysis.

油气评价分析基于不同层次的研究，如盆地、沉积序列和油气系统分析研究等。

evaluation of oil and gas area /district evaluation *np.* 区带评价

例：There are mainly 4 key parameters of natural gas resource evaluation, including gas production rate, migration-accumulation coefficient, evaluation of oil and gas area, and recovery rate.

天然气资源评价主要包括产气率、运聚系数、油气区块面积评价与采收率 4 个重点参数。

trap assessment/evaluation *np.* 圈闭评价

例：In the practical operation of oil and gas exploration, the trap assessment is multi-sample, only limited data are obtained.

在油气勘探实际操作中，圈闭评价形式多样，能获取的资料有限。

hydrocarbon reservoir evaluation *np.* 油气储层评价

例：The permeability of the formation is key parameter which can demonstrate the complexity of reservoir pore structure and effect the dynamic prediction. So permeability directly influence the hydrocarbon reservoir evaluation.

地层渗透率是反映储层孔隙结构复杂程度的重要参数，影响动态预测。因此，渗透率对油气储层评价有直接影响。

reservoir characterization *np.* 油藏特性描述

例：Better reservoir characterization will help reduce risk during the stage of appraisal, development and into production.

油藏特性描述得好有助于降低评价、开发乃至投产阶段的风险。

reservoir monitoring *np.* 油藏监控

例：Reservoir monitoring through time-lapse seismic can play an invaluable role in reducing risk and maximizing production.

通过时差地震进行油藏监控将在降低风险和提高产量方面起重要作用。

streamer technology *np.* 漂缆技术

例：In many areas, streamer technology will suffice to provide the necessary information throughout the life of the oil field.

在许多领域，漂缆技术足以在油田的整个生命周期内提供必要的信息。

formation evaluation *np.* 地层评价

例：Reservoir parameters are the bases for formation evaluation.
储层参数是地层评价的基础。

gas bearing evaluation *np.* 含气性评价

例：Above the base statement, scientists comment the gas bearing evaluation and point out the helpful the stratum of containing gas and area.
在此基础上，科学家们进行了含气性评价，指出了有利的含气层位及区域。

water cut *np.* 含水率

例：Economy limit water cut is an important decision making gist to oil field exploitation.
经济极限含水率是油气田开发决策的重要参考依据。

oil content *np.* 含油量

例：Total oil content is the sum of the dispersed and dissolved oil content of an oil containing mixture.
总含油量是含油混合物的分散油含量和溶解油含量之和。

residual oil saturation *np.* 残余油饱和度

例：We estimate original oil in place, residual oil saturation, and remaining mobile oil.
我们估算原油地质储量、残余油饱和度和剩余可动油量。

gross lithology *np.* 总岩性

例：In the early stages of exploration certain general conclusions as to the distribution and quality of potential reservoir could be made from their gross lithology.
在勘探早期，可以从总岩性中得出一些关于潜在储层分布和质量的一般性结论。

oil-bearing area *np.* 含油区

例：A number of oil/ gas pools and a large oil-bearing area constitute a complex hydrocarbon accumulation belt favorable for oil-gas exploration and development.
很多油气藏和大型含油区形成了有利于油气勘探开发的复杂油气聚集带。

pure oil area *np.* 纯油区

例：The oil wells that are located in the pure oil areas with distance less than 200 m from edge water must properly avoid being perforated.

油井位于纯油区，且离边水距离不足 200 米时，要适当考虑避免射孔。

oil/water transition zone *np.* 油水过渡带

例：The calculation of fluid saturation in the transition zone is highly related to accurate calculation of oil and gas geologic reserve in oil/ water transition zone in heterogeneous low permeability reservoirs.

确定油藏过渡带的流体饱和度，是准确计算非均质、低渗透油藏中油水过渡带油气地质储量的关键。

fluid contact *np.* 流体界面

例：Fluid contacts in a trap are generally planar, but are by no means always horizontal.

圈闭中的流体界面通常是平面的，但决不总是水平的。

oil–water contact (OWC) /oil/water interface *np.* 油水界面

例：The oil-water contact (OWC) is the deepest level of producible oil.

油水界面（OWC）是可采原油的最深部位。

gas/oil contact *np.* 气 / 油接触面

例：Any pressure decline due to oil production results in an expansion of the gas cap, and the gas/oil contact moves downwards.

石油产出导致压力下降，气顶膨胀，气 / 油接触面向下移动。

associated gas cap *np.* 伴生气顶

例：Since it provides important reservoir energy, an associated gas cap is not produced until all the recoverable oil is produced from the reservoir.

伴生气顶提供了重要的储层能量，因此只有全部可采石油从储层中采出，伴生气顶气才能采。

simulated maturation *np.* 模拟熟化

例：Qualitative information can be obtained from laboratory "simulated maturation" experiments.

从实验室的"模拟熟化"实验中可获得定性信息。

vitrinite reflectance *np.* 镜质体反射率

例：Vitrinite reflectance is a method used to determine the maturity of a source rock.

镜质体反射率是用于确定烃源岩成熟度的方法。

hydrocarbon detection *np.* 油气检测

例：Hydrocarbon detection technique is a technical method that comprehensively uses the physical and chemical information related to oil and gas content to determine the enrichment zone of oil and gas.

油气检测技术是一种综合利用与油气含量相关的物理、化学信息来确定油气富集带的方法。

geochemical indicator *np.* 化探指标

例：The anomaly areas of different indicators are well superposed and they are slightly larger than the area of gas reservoir, which shows that the geochemical indicator combination is effective.

不同指标的异常范围叠合好，异常范围略大于气藏的面积，说明化探指标组合有效。

formation factor *np.* 地层因素

例：AFF (acoustic formation factor) equation is an effective porosity interpretation model using sonic logging data.

AFF（声波地层因素）方程是利用声波时差资料求取地层孔隙度的有效解释模型。

formation water *np.* 地层水

例：Although formation water normally is the same as the geological formation water, or interstitial water, it may be different because of the influx of injection water.

虽然地层水通常与地质地层水或间隙水相同，但也可能因注入水涌入而不同。

reservoir pressure *np.* 储层压力

例：Reservoir pressure during conversion and steam drive must be controlled below 5MPa.

转驱及蒸汽驱过程中的储层压力必须控制在 5 兆帕以下。

original solution gas/oil ratio *np.* 原始溶解气油比

例：Original solution gas/oil ratio, formation permeability, pressure drop rate are all sensitive to well deliverability except shape factor.

除了形状因子，原始溶解气油比、地层渗透率、压降率对油井产能都很敏感。

logging evaluation *np.* 测井评价

例：Array acoustic wave data are widely applied in well logging evaluation due to its abundant stratigraphic information.

由于能提供丰富的地层信息，阵列声波数据在测井评价中得以广泛应用。

well testing *np.* 试井

例：The accurate evaluation of these surfaces is essential before the reserves of a field can be calculated, and their establishment is one of the main objectives of well logging and testing.

在计算油田储量之前，必须对这些界面进行准确评估，而确定和建立这些界面是测井和试井的主要目标之一。

productivity predication *np.* 产能预测

例：Horizontal well productivity predication and evaluation are important issues at its design stage for a gas condensate reservoir.

水平井产能预测与评价是凝析气藏水平井设计中的重要方面。

pay *n.* 产层

例：Within the trap the productive reservoir is termed the pay.

圈闭内的可产储层称为产层。

gross pay *np.* 总产层

例：The vertical distance from the top of the reservoir to the petroleum-water contact

is termed gross pay.
从储层顶部到油水界面的垂直距离称为总产层。

net pay *np.* 净产层

例：All of the gross pay does not necessarily consist of productive reservoir, so gross pay is usually differentiated from net pay.
并非总产层都包括生产性储层。因此，总产层有别于净产层。

reserve *n.* 储量

例：The longer the history the better the estimates of reserves.
开采历史越长，对储量估计就越准确。

reserve evaluation *np.* 储量评价

例：Reserve evaluation is the basement of researches on the feasibility of oil and gas field development.
储量评价是油气田开发可行性研究的基础。

predicted reserve *np.* 预测储量

例：These achievements have been widely used in the appraisal of controllable predicted reserve, laying a good foundation for future well placement and reserve arrangement.
在评价可控预测储量中，这些成果得到了广泛的应用，为今后的井位布置和储量安排奠定了良好的基础。

volume of in-place petroleum *np.* 石油地质储量

例：If the volume of in-place petroleum is unavailable, recoverable hydrocarbons are the next best volume.
如果无法获得石油地质储量，也可以用可采储量来比较。

initial oil in place/original oil in place (OOIP) *np.* 石油地质储量

例：Underground compressed correction of porosity is of great importance for the OOIP calculation.
地下压缩岩孔隙度校正对计算石油地质储量有重要的意义。

proved reserve *np.* 探明储量

例：The prediction of incremental proved reserves is important for decision making in petroleum exploration.

增量探明储量预测对石油勘探决策具有重要意义。

recoverable reserve *np.* 可采储量

例：The recovery factor also may be changed in the light of subsequent production performance, or as a result of the introduction of secondary recovery techniques, which will result in a revision of the proven recoverable reserves figure.

根据后续生产及二次采油技术的实施情况，采收率也可能会发生变化，从而已探明可采储量也可能有变化。

reserves abundance *np.* 储量丰度

例：Permian-Triassic fans have plenty of oil-bearing formation, large scale of petroleum pool, higher reserves abundance.

二叠与三叠纪扇体地层含油层丰富、油藏规模大、储量丰度高。

large oil and gas fields *np.* 大型油气田

例：The quantity of oil and gas resources in a certain plateau is abundant and it is likely that large oil and gas fields may be found in the plateau basin.

高原盆地的油气资源量较为丰富，可找到大型油气田。

middle oil–gas field *np.* 中型油气田

例：The latest resources assessment result indicated that oil-gas resources is plentiful where total petroleum resources is 2.7–3.2 hundred million tons and that of nature gas is $(1.75-2.94) \times 10^4$ hundred million cubic meter. It shows that there can form middle or large oil-gas field.

最新资源评价结果表明油气资源丰富，石油总资源量为 2.7 亿 ~ 3.2 亿吨，天然气总资源量为 1.75 万亿 ~ 2.94 万亿立方米，该区具备形成大中型油气田的资源基础。

Part II

Oil and Gas Drilling Engineering

油气井工程

Section I

Wells Categories and Operators
油气井类型及作业人员

cable tool drilling *np.* 顿钻钻井

例：Advances in well completion practices have brought to rotary drilling many of the advantages for cable tool drilling.

完井作业的诸多技术进展给旋转钻井带来很多优势，这是顿钻钻井所不可比拟的。

rotary drilling *np.* 旋转钻井

例：Modern rotary drilling requires careful manipulation of the drilling fluid in order to reach target depth successfully.

现代旋转钻井需要精心配制钻井液，以成功达到目标深度。

horizontal drilling *np.* 水平钻井

例：Advances in horizontal drilling have improved drilling efficiency and gas production rates.

水平钻井技术的进步提高了钻井效率和页岩气开采效率。

directional drilling *np.* 定向钻井

例：The key problem of directional drilling is the accuracy and reliability of orientation of deflecting tools.

定向钻井的关键在于造斜工具定向的准确性和可靠性。

air drilling *np.* 空气钻井

例：The air screw drill tool was power equipment which was designed for air drilling.

空气螺杆钻具是专为空气钻井而设计的动力装置。

balanced pressure drilling *np.* 平衡压力钻井

例：Theoretically, there is no pressure difference in <u>balanced pressure drilling</u>, but in practice it is impossible to achieve complete pressure balance.

理论上讲，平衡压力钻井不存在压力差，而实际上要做到完全压力平衡是不可能的。

under–balanced drilling *np.* 欠平衡压力钻井

例：Air drilling is a kind of <u>under-balanced drilling</u> technology with air, whose rock breaking form is different from conventional mud drilling.

空气钻井是一种以空气作为循环介质的欠平衡压力钻井技术，其破岩形式与常规钻井液钻井有很大区别。

top drive drilling *np.* 顶驱钻井

例：Currently, the research of <u>top drive drilling</u> system has been matured, but still needs developing for its reliability.

目前顶驱钻井技术研究基本成熟，但其可靠性研究有待进一步提高。

floating marine operation *np.* 海上浮式钻井船作业

例：In <u>floating marine operations</u> the hazards are much greater than one on land or fixed platform well.

海上浮式钻井作业风险远大于陆上或固定平台钻井作业。

exploratory drilling *np.* 勘探钻井

例：Although many famous oil fields were discovered by guess and by gosh, the fact remains that the only sure way to prove that oil or gas lies buried under some likely spot is to probe for it with a <u>exploratory drilling</u>.

尽管许多大油田都是凭猜测发现的，但事实上要证明某地可能埋藏有石油或天然气，唯一可靠的方法就是通过勘探钻井进行探测。

straight well *np.* 直井

例：If the survey calculation takes place on the <u>straight well</u> path, the model error of survey calculation could be ignored.

如果对直井轨迹进行测量计算，模型误差可予以忽略。

horizontal well *np.* 水平井

例：Drilling a horizontal well is more difficult than a vertical well.

水平井比垂直井更难钻。

directional well *np.* 定向井

例：It is beneficial to carry out the construction of directional well through the analysis of directional differential pipe-sticking.

对定向钻井压差卡钻进行分析，有利于更好地进行定向井作业。

sidetrack well *np.* 侧钻井

例：Some new logging problems have appeared because horizontal, extended reach and sidetrack wells have been widely drilled in recent years.

近年来由于水平井、大位移井和侧钻井的广泛应用，出现了一些新的测井问题。

sidetrack horizontal well *np.* 侧钻水平井

例：Sidetrack horizontal well can increase oil production, extend the production life of injecting water reservoirs and enhance oil recovery.

侧钻水平井可提高原油产量，延长注水油藏的生产寿命，提高采收率。

high angle well *np.* 大斜度井

例：Well trajectory control is one of the key technologies of high angle well drilling, and the results of the control directly affect the well friction, torque, safety and construction of all wells.

井眼轨迹控制是大斜度井关键技术之一，其结果直接影响该井摩阻、扭矩、全井的安全及施工等。

cluster well *np.* 丛式井

例：The rotary drill stem will engender direction drift in rotary drilling of directional well and cluster well.

定向井及丛式井在旋转钻井作业时，旋转钻柱会发生方位漂移。

multilateral well *np.* 多分支井

例：Multilateral well technology can only be utilized when the driver for the technology is understood.

只有知晓多分支井技术动因，才能使用多分支井技术。

lateral *n.* 分支井

例：Complete isolation of a drilled and/or completed lateral while working on another lateral is essential to prevent well control problems.

为了防止井控问题发生，在进行另一个分支井作业时，必须对已钻和 / 或已完成的分支井进行完全封闭。

multi–bore well *np.* 多底井

例：The gel cement has better embrittlement resistance and little shrinkage, and can provide long time supporting for the liner, thus it is an ideal select for multi-bore well cementing.

乳胶水泥抗脆裂性好、收缩性小，能对尾管提供长期支撑，是多底井固井理想的水泥浆体系。

infill well *np.* 加密井

例：With the increase of infill wells and sidetrack wells, more and more sidetrack wells require more bypassing obstacles and become more difficult to be planned and drilled.

随着加密井和侧钻井数量不断增加，越来越多侧钻井需要采取绕障措施，其设计和施工难度也越来越大。

relief well *np.* 救援井 / 减压井

例：The relief well is a directional well drilled at a safe distance on the surface from the wild well.

救援井是在地面与野喷井 / 失控井相距一个安全距离处所钻的一口定向井。

wildcat well /exploration well *np.* 初探井 / 勘探井

例：Although it is certainly better than a blind guess or a hunch, the odds are that six of every seven new, or wildcat wells drilled will be dry.

钻勘探井当然比盲目猜测好，但新钻的勘探井中每七口可能有六口都是干井。

step-out well *np.* 探边井

例：Decisions will be taken and the location chosen for the drilling of appraisal, or step-out wells ascertains the extent of the field.

需要做出决定，并选择评价井或探边井的井位，以确定油田范围。

appraisal well /step-out well *np.* 评价井

例：First comes the decision to drill appraisal wells, to determine more reliably the size of the accumulation, and to test the reservoirs for production capacity and quality.

首先要决策评价井钻探，确定储层规模，并测试储层的产能和质量。

development well *np.* 开发井

例：A development well is a well that is drilled after an exploration well has confirmed the presence of petroleum in the formation.

开发井是在探井证实地层中有石油后钻探的井。

producer *n.* 生产井

例：If the well has been successful, and is a potential producer, a method of "completing" the well must be chosen.

如果钻井成功，且可能产油，则必须选择"完井"方法。

dry/barren well *np.* 干井

例：If the decision is to abandon it, the hole is considered to be a dry well not capable of producing oil or gas in commercial quantities.

如果决定放弃该井，该井则被视为无商业价值油气产量的干井。

abandoned well/well abandonment *np.* 弃井

例：The abandoned wells that were plugged as dry at one time in the past may be reopened and produced if the price of oil or gas has become more favorable.

如果石油或天然气价格变得更为有利，过去一度堵塞废弃的干井可能会重新开放生产。

rig crew *np.* 钻井队

例：The rig crew usually assists in casing and cementing operations.

钻井队通常协助套管和固井作业。

casing crew *np.* 套管队

例：Special <u>casing crews</u> are hired to run the casing and usually a cementing company is called on to place cement around the casing to bond it in place in the hole.

雇佣专业套管队来下套管，通常需要固井公司在套管周围放置水泥，将套管固定在井中。

cementing service company *np.* 固井服务公司

例：An oil well <u>cementing service company</u> is usually called in for this job, although, as when casing is run, the rig crew is available to lend assistance.

这项工作通常需要一家油井固井服务公司，下套管时钻井队可以提供帮助。

logging crew *np.* 测井队

例：To lower or hoist the cable, the <u>logging crew</u> has to apply a drill rig.

要降低或提升电缆，测井人员必须使用钻机。

service company *np.* 修井公司

例：<u>Service companies</u> have calibration procedures for most tools, some of which are based on standards established by the American Petroleum Institute (API).

修井公司大多数工具都有校准程序，有些校准程序是基于美国石油协会（API）制定的。

tool pusher *np.* 钻井队长

例：If we had bits that would not wear out, or that would drill perfectly no matter how operated, we would not have to have drillers and <u>tool pushers</u>.

如果钻头不会磨损，或者无论如何操作都能完美地钻进，就不需要司钻和钻井队长了。

derrickman *n.* 井架工

例：The <u>derrickman</u> works above the rig floor, near the top of the derrick, where he attaches or detaches the elevators when pipe or casing is run into or pulled out of the hole.

井架工在钻台上方靠近井架顶部工作，在起下钻杆或套管过程中，井架工扣上或卸下吊卡。

driller *n.* 司钻

例：The change of the penetration rate and the behavior of the rotary table can give the driller evident hint that there is a change of formation at the depth being drilled.

钻速发生变化和转盘旋转情况都可以给司钻明显信号，在钻地层发生了变化。

roughneck/floorman *n.* 钻工

例：Among a roughneck's duties are such things as operating the cat-head, handling the slips and tongs, standing pipe back in the derrick, assisting in mixing the slush, and so on.

钻工的职责包括操作猫头，用卡瓦，打大钳、排立根，帮助搅拌钻井液等。

roustabout *n.* 杂工

例：A roustabout does semi-skilled labor such as scraping rust hosing down, painting, carrying cans of dope, unloading materials and supplies, etc.

杂工从事半技术性工作，如刮锈、冲洗、喷漆、搬运涂料罐、卸载物品等。

floor crew *np.* 钻台工作队

例：The floor crew sets the slips and swings two big wrenches called tongs into action.

钻台工作队安放好卡瓦，然后操作大钳。

rotary crew *np.* 钻工班

例：Even if a man has a university or a polytechnic education, most oil companies will want him to get rig experience by working on the floor with the rotary crew for a certain period.

即使接受过大学教育或职业教育，大多数石油公司仍要求这些员工和钻井班一起在钻井台工作一段时间，积累钻井工作经验。

rig supervisor *np.* 井场监督

例：A fishing tool operator can be expected to do a better job than a rig supervisor who has to run a specific tool only rarely, and who usually has had no experience with the new tools and methods.

专业打捞人员更在行，井场监督不常用专业打捞工具，对工具及其使用方法都不熟。

workover engineer *np.* 修井工程师

例：Fishing technique in horizontal well has always been the tackling key problem for <u>workover engineers</u> and it is also the difficulty in workover job.

水平井打捞技术一直是修井工程师们努力攻关的方向，也是修井作业的难点。

well puller *np.* 修井工

例：There are certain skills that many <u>well pullers</u> have in order to accomplish their responsibilities.

很多修井工都具备一定的技能以履行其工作职责。

mud logger *np.* 钻井液录井员

例：The <u>mud logger</u> catches cuttings at the shale shaker and by using a microscope or ultraviolet light can see whether oil is in the cuttings.

钻井液录井员在振动筛处获取岩屑，通过显微镜或紫外线可以看到岩屑中是否有石油。

well logger *np.* 测井员

例：Using a portable laboratory, truck-mounted for land rigs and permanently mounted on offshore rigs, the <u>well loggers</u> lower devices called logging tools into the well on wireline.

海上测井把便携式实验室固定安装在钻井平台上，测井员通过电缆将测井仪下放到油井中。

field engineer *np.* 油田工程师

例：<u>Field engineers</u> are to supervise drilling, completion, and well workover operations.

油田工程师监管钻井、完井和修井作业。

Section II

Drilling Equipment
钻井设备

rotary rig *np.* 旋转钻机

例：The main function of a rotary rig is to drill a hole.
旋转钻机的主要功能是钻井眼。

prime mover *np.* 原动机

例：Practically every rig uses internal-combustion engines as its prime power source, or its prime mover.
每台钻机都使用内燃机作为其原动力源或原动机。

generator *n.* 发电机

例：Automatic voltage regulator can ensure a constant output from the generator.
自动电压调节器可保证发电机电流输出稳定。

diesel engine *np.* 柴油发动机

例：Diesel engines are designed to be more efficient than gasoline engines, so they provide higher fuel efficiency.
柴油发动机设计得比汽油发动机效率高，因此可以提供更高的燃油效率。

derrick *n.* 井架

例：Strictly speaking, a derrick is a towerlike structure that must be assembled piece by piece, requiring a rig-building crew to fasten the pieces together with bolts.
严格地说，井架是一种塔式结构，必须一件一件地组装，需要钻井施工人员用螺栓将这些部件固定在一起。

mast *n.* 桅杆式井架

例：A mast can be raised or lowered without disassembly.

桅杆式井架可以在不拆卸的情况下整体起放。

hoisting system *np.* 提升系统

例：The hoisting system is made up of the drawworks (sometimes called the hoist), a mast or derrick, the crown block, the traveling block, and wire rope.

提升系统由绞车（有时称为起重机）、桅杆或井架、天车、游车和钢丝绳组成。

crown block *np.* 天车

例：The number of sheaves needed on the crown block is always one more than the number needed in the traveling block.

天车滑轮通常比游车的多一个。

pulley blocks *np.* 滑轮组

例：The derrickman also cleans, oils, greases, inspects and repairs the pulley blocks and cables which are used to raise and lower sections of pipe and casing.

井架工还要负责清洁、润滑、检查和修理滑轮组和电缆，滑轮组和电缆用于升降钻杆和套管。

traveling block *np.* 游车

例：Attachments to the traveling block include a large hook to which the equipment for suspending the drill string is attached.

游车下面接着大钩，大钩上附有悬挂钻柱的设备。

drawwork *n.* 绞车

例：The primary function of the drawworks is to reel out and reel in the drilling line, a large diameter wire rope, in a controlled fashion.

绞车的主要功能是控制钻井绳卷出和卷入，钻井绳为大直径钢丝绳。

drilling line *np.* 钻井绳

例：For wire rope to be useful as the drilling line, it has to be strung up, for it arrives at the rig wrapped on a large supply reel.

钢丝绳用作钻井绳必须将其串起来，因为钢丝绳到达钻机时缠绕在一个大型供带盘上。

drawworks drum *np.* 绞车滚筒

例：To lower the traveling block, line is let out of the drawworks drum.

为了降低游车，要将钢丝绳从绞车滚筒中放出。

swivel *n.* 旋转接头（水龙头）

例：Rotating equipment from top to bottom consists of a wondrous device known as the swivel, a short piece of pipe called the kelly, the rotary table, the drill string, and the bit.

旋转设备自上而下分别为水龙头（一种特殊装置）、方钻杆（一截短管）、转盘、钻柱和钻头。

bail *n.* 提环

例：The swivel also has a large bail, similar to the bail or handle on a bucket but much, much larger, which fits inside the hook at the bottom of the traveling block.

水龙头有一个大的提环，类似于桶上的提环或手柄，但要大得多，安装在游车底部的挂钩内。

hook *n.* 吊钩

例：The hooks are used to hang various other equipment, particularly the swivel and kelly, the elevator bails or top drive units.

吊钩用于悬挂各种其他设备，主要有转盘、方钻杆、升降机吊环或顶驱装置。

rotary hose *np.* 水龙带

例：The mud is pumped up the drill pipe and into the rotary hose, goes down the drill collars and exits at the bit.

钻井液被泵入钻杆并进入水龙带，沿钻铤向下并在钻头处排出。

kelly (bar/rod) *n.* 方钻杆

例：Immediately below the swivel is attached a square (four-sided) or hexagonal (six-sided) piece of pipe called the kelly.

在旋转接头的正下方，连接着一根称为方钻杆的方形（四边形）或六边形（六边形）钻杆。

kelly bushing *np.* 方钻杆补心（方补心）

例：The kelly fits inside a corresponding square or hexagonal opening in a device called a <u>kelly bushing</u>, which is a part of the rotary table.

方钻杆置于相应的四边形或六边形的方补心内，方补心为转盘的一部分。

master bushing *np.* 主补心

例：The kelly bushing, in turn, fits into another part of the rotary table called the <u>master bushing</u>.

方补心又安装到转盘的另一部分里，称为主补心。

slip *n.* 卡瓦

例：The master bushing drives the kelly bushing and accommodates a device called the <u>slips</u>.

主补心驱动方补心，其中还有一个称为卡瓦的装置。

rotary table *np.* 转盘

例：Powered by its own electric motor, the <u>rotary table</u> is comprised of several parts.

转盘由几个部分组成，由自身电动机驱动。

drill pipe *np.* 钻杆

例：The measured length of each joint of <u>drill pipe</u> or tubing is added to provide a total depth or measurement to the point of interest.

测量每节钻杆或油管的长度，并将其相加，便可提供目的位置的总深度。

tool joint *np.* 钻杆接头

例：The torsional strength of the drilling tools can be greatly increased by using the double shoulder <u>tool joint</u>.

双台肩钻杆接头可大幅度提高钻具的抗扭强度。

drill collar *np.* 钻铤

例：<u>Drill collars</u> are heavier than drill pipe and are used on the bottom part of the string to put weight on the bit.

钻铤比钻杆重，用于钻柱底部，给钻头加重。

drill stem *np.* 钻柱

例：Officially, the assembly of members between the swivel and the bit, including the kelly, drill pipe, and drill collars, is termed the drill stem.

水龙头至钻头这部分组件总体正式称为钻柱，两者之间包括方钻杆、钻杆和钻铤。

drill string *np.* 钻柱

例：The term drill string refers simply to the drill pipe; however, in the oil patch drill string is consistently used to mean the whole works.

钻柱这一术语仅指钻杆，然而，在油田现场，钻柱用于表示整个组件（类似 drill stem）。

drill bit *np.* 钻头

例：The drill bit is on the bottom of the drill string and must be changed when it becomes excessively dull.

钻头位于钻柱末端，磨损严重时必须更换。

drag bit *np.* 刮刀钻头

例：Diamond bits function the same as drag bits, particularly in the fact that both weight and rotary speed are directly related to drilling speed.

金刚石钻头的功能与刮刀钻头相同，尤其是重量和转速都与钻速直接相关。

diamond bit *np.* 金刚石钻头

例：A new special diamond bit combines ultrasonic vibrating with grinding of diamond grits.

一种新型金刚石钻头把超声振动和金刚石磨粒磨削结合在一起。

Polycrystalline Diamond Compact bit (PDC) *np.* 聚晶金刚石复合片钻头

例：It is the key to ensure that PDC bit has the choice of the work performance to reasonably place cutting.

合理布齿是保证 PDC 钻头优良工作性能的关键。

cone bit *np.* 牙轮钻头

例：The complex layers are compacted and well-On the system dynamics, the dynamics model for roller cone bit under the interaction of drilling string and rock was established.

从系统动力学的角度出发，建立基于钻柱、岩石相互作用下的牙轮钻头动力学模型。

jet bit *np.* 喷射式钻头

例：Jet bits have nozzles that direct a high velocity stream or jet of drilling fluid to the sides and bottom of each cone, so that rock cuttings are swept out of the way as the bit drills.

喷射式钻头装有喷嘴，钻井液从喷嘴喷出形成高速旋流，随着钻头的转动把井底或边缘的岩屑很快地冲离钻头。

weight indicator *np.* 指重表

例：Using a special instrument—a weight indicator—the driller monitors the amount of weight put on the bit by the drill collars.

司钻使用一种特殊仪器（指重表），监测钻铤加在钻头上的钻压。

rotary steerable systems (RSS) *np.* 旋转转向系统

例：Rotary steerable systems (RSS), categorized into two types: "push-the-bit" and "point-the-bit" systems, have enabled directional drilling to become both more effective.

旋转转向系统（RSS）分为两类："推式钻头"和"摆动钻头"，使定向钻井更加经济、有效。

push-the-bit *np.* 推靠式钻头

例：Push-the-bit systems incorporate bias and control units into the BHA.

推靠式钻头系统在井底钻具组合中装有偏置器和控制器。

point-the-bit *np.* 指向式钻头

例：Point-the-bit systems steer the bit by tilting it in the direction of the desired angle.

指向式钻头系统可使钻头按预定方向倾斜以控制钻头方向。

steerable motor *np.* 转向马达

例：A steerable motor is attached to the drill string behind the drill bit.

转向马达安装在钻头后面的钻柱上。

bottom hole assembly (BHA) *np.* 井底钻具组合

例：The bottom hole assembly (BHA) is comprised of the drill bit, collars to apply weight to the drill bit, and stabilizers to keep the drilling assembly centered in the borehole.

井底钻具组合(BHA)由钻头、给钻头加压的钻铤以及保持钻具组合处于井眼中心的平衡器组成。

elevator *n.* 吊卡

例：It is suggested that taper elevator shoulder be used to prevent the crack of tool joint in essential.

要从根本上防止钻杆接头断裂，建议采用斜坡吊卡台肩钻杆。

brake *n.* 制动器

例：The brake is the mechanism on the drawworks that permits the driller to control the speed and motion of the drill string.

司钻可用绞车制动器控制钻柱速度和起停。

mud pit *np.* 钻井液池

例：It is very important to measure the volume change of mud in a mud pit for the safe drilling, the control of well blowout and the loss of circulation.

测量钻井液池中钻井液体积的变化对于安全钻井，控制井喷、井漏具有十分重要的意义。

jet–mixing hopper *np.* 漏斗混合器

例：Jet-mixing hopper is the equipment to match with the drilling solids control equipments, regularly used to mix and adjust the proportion of drilling fluid, changing the density and viscosity etc.

漏斗混合器与钻井固控设备配套，用于定期混合和调整钻井液比例、改变其密度和黏度等。

mud agitator *np.* 搅拌器

例：Mud agitators are needed to maintain mud weight material in suspension.
钻井液搅拌器用来保持钻井液重质颗粒处于悬浮状态。

shale shaker *np.* 振动筛

例：Desilters and desanders are installed on the mud pits to remove very small particles that can go through the shale shaker.
除泥器和除砂器安装在钻井液池上，可以去除通过振动筛的较小颗粒。

mud pump *np.* 钻井泵

例：The diaphragm pump is a kind of updating product for the traditional mud pump, whose heart component is a crankshaft.
隔膜泵是传统钻井泵的更新换代产品，曲轴则是隔膜泵的核心部件。

reciprocating pump *np.* 往复泵

例：Reciprocating pumps use a piston moving back and forth in a cylinder.
往复泵通过活塞在气缸中来回移动。

shearometer *n.* 钻井液静切力计

例：Shearometer is an instrument used together with a set of weights to conduct a shear-strength measurement test.
钻井液静切力计是一种配合砝码进行剪切力强度测试的仪器。

mud centrifuge *np.* 钻井液离心机

例：Mud centrifuge is a special equipment designed for solid-liquid separation according to the characteristics of petroleum drilling fluid.
钻井液离心机是针对钻井液特点而设计的固液分离专用设备。

centrifugal pump *np.* 离心泵

例：Centrifugal pump is widely used in power plants, oil, chemical industry, urban water supply, agricultural irrigation and drainage sectors.
离心泵广泛用于发电厂、石油、化工、城市供水、农业排灌等行业。

mud cleaner *np.* 钻井液清洁器

例：The mud cleaner is the primary and secondary purification equipment in the solid control system of the drilling mud.

钻井液清洁器是钻井液固控系统中的一级和二级净化设备。

desilter *n.* 除泥器

例：A sand pump is used to provide drilling liquid with a certain discharge capacity and pressure to desander, <u>desilter</u> and mud pump.

砂泵主要用来为除砂器、除泥器、钻井泵提供具有一定排量和压力的钻井液。

degasser *n.* 除气器

例：If the gas content in the mud is high, a mud gas separator is used, because it has a higher capacity than standard <u>degassers</u>.

如果钻井液中的气体含量很高，则使用钻井液气体分离器，因为它比标准除气器具有更高的容量。

desander *n.* 除砂器

例：The <u>desander</u> should be located downstream of the shale shakers and degassers, but before the desilters or mud cleaners.

除砂器应位于振动筛和除气器下游，以及除泥器或钻井液清洁器上游。

mud-gas separator *np.* 钻井液气体分离器

例：<u>Mud-gas separator</u> discharges gas and transfers mud to the storage tank.

钻井液气体分离器排出气体并将钻井液输送至钻井液罐。

mud pit instrumentation *np.* 钻井液池检测仪

例：<u>Mud pit instrumentation</u> has the functions of temperature setting, fine tuning, temperature compensation and automatic control.

钻井液池检测仪具有温度设定、温度微调、温度补偿及自动控制等功能。

reserve pit *np.* 钻井液储备池

例：This <u>reserve pit</u> serves as a place of disposal for used or unneeded drilling mud, cuttings from the hole, and other waste that invariably accumulates around the site.

钻井液储备池用于处置废弃钻屑、钻井液和现场周围堆积的废物。

well–control equipment *np.* 井控设备

例：During drilling, it is very important to have a reliable well-control equipment in order to prevent welling up and blowing and to realize safe and effective exploration and development.

在钻井作业中，为了防止井涌和井喷，实现安全高效的勘探开发，可靠的井控设备十分重要。

shutoff device *np.* 关闭装置

例：There was no shutoff device, and the well produced from wide-open casing for nine days before a valve could be attached and closed to stem the flow.

该井没有关闭装置，在套管全开情况下生产了 9 天，才装阀门关井，阻止流体流动。

wellhead *n.* 井口

例：The positive choke valve is employed on Christmas trees and wellheads and designed for flow control.

固定油嘴用于采油树和井口以控制流量。

blowout preventer *np.* 防喷器

例：After the cement hardens and tests indicate that the job is good, the rig crew attaches or nipples up the blowout preventer stack to the top of the casing.

水泥凝固后，经测试表明固井质量达标，钻井队才将防喷器安装在套管顶部。

annular preventer *np.* 环形防喷器

例：The annular preventer is usually mounted at the very top of the stack of BOPs.

环形防喷器通常安装在防喷器的最顶部。

ram–type preventer *np.* 闸板式防喷器

例：Usually, only the annular preventer will be closed when the well kicks, but should it fail, or should it be necessary to use special techniques, the ram-type preventers are used as a backup.

井涌时，通常只关闭环形防喷器，但如果失灵，或需使用特殊方法，则启用闸板式防喷器。

double-ram-type preventer　*np.* 双闸板防喷器

例：The double-ram-type preventer consists of two pairs of rams which are operated by independent controls, normally hydraulic.

双闸板防喷器由两对独立控制（通常为液压控制）的闸板组成。

sleeve type preventer　*np.* 套筒式防喷器

例：The sleeve-type blowout preventer (Hydril) is at the top of the service wellhead.

套筒式防喷器（旋转防喷器）安装于修井井口装置的顶部。

platform　*n.* 钻井平台

例：Mobile offshore rigs require less rig-up time than platforms because most of the equipment is already in place and assembled.

移动式海上钻机的安装时间比钻井平台少，因为大多数设备已经就位和组装完。

semisubmersible　*n.* 半潜式平台

例：Floaters, such as semisubmersibles and drill ships, simply have to be anchored on location, and drilling operations can begin.

浮式钻井平台，如半潜式平台和钻井船，只需固定到位，即可开始钻井作业。

floater　*n.* 移动式钻井平台

例：Some floaters are not anchored; instead, dynamic positioning is utilized to keep the rig on station (in a position more or less directly over the spot on the seafloor where the hole is to be drilled).

有些移动式钻井平台不用抛锚；通过动态定位使钻机稳定于海底预定井位上方。

jack-up　*np.* 自升式钻塔

例：The increasing demand for oil has led to deeper drilling, larger structures located further offshore and the development of "jack-up" and semi-submersible drill rigs with greater capabilities.

石油需求不断增长，促使钻井更深、平台更大、离岸更远，并开发出高性能自升式钻塔和半潜式钻机。

Section III

Drilling Fluid
钻井液

drilling fluid/mud *np.* 钻井液

例：<u>Drilling fluid/mud</u>—is usually a mixture of water, clay, weighting material, and a few chemicals.

钻井液通常将水、黏土、加重材料和一些化学品混合制成。

water–based mud (fluid) *np.* 水基钻井液

例：Oil-based mud are generally more expensive and require more stringent pollution control procedures than <u>water-based mud</u>.

油基钻井液通常比水基钻井液昂贵，需要更严格的污染控制程序。

oil–based mud/fluid *np.* 油基钻井液

例：<u>Oil-based mud</u> are sometimes used to avoid the damaging influence on well productivity by aqueous filtrate.

有时会使用油基钻井液以避免滤失对油井产能造成破坏。

synthetic–based drilling fluid *np.* 合成基钻井液

例：Diluent is added to <u>synthetic-based drilling fluid</u> as rheological regulator.

合成基钻井液加入稀释剂可调节钻井液的流变性。

seawater based fluid *np.* 海水钻井液

例：Obviously, there is a great advantage in using <u>seawater based fluids</u> for most offshore operations.

显然，大多数海上作业中使用海水钻井液具有很大的优势。

saltwater drilling fluid *np.* 盐水钻井液

例：Drilling fluids with NaCl content exceeding 1% are collectively referred to as saltwater drilling fluid.

凡氯化钠（NaCl）含量超过1%的钻井液统称为盐水钻井液。

calcium treated drilling fluid *np.* 钙处理钻井液

例：Calcium treated drilling fluid inhibits the hydration and dispersion of clay.

钙处理钻井液可以抑制黏土的水化分散作用。

polymer drilling fluid *np.* 聚合物钻井液

例：Polymer drilling fluid helps to keep the borehole wall stable.

聚合物钻井液有利于保持井壁稳定。

dispersed drilling fluid *np.* 分散钻井液

例：Dispersed drilling fluid is easy to form a dense mud cake on the well wall.

分散钻井液容易在井壁上形成较致密的滤饼。

low solids drilling fluid *np.* 低固相钻井液

例：Low solids drilling fluid can significantly improve ROP.

低固相钻井液可大幅提高机械钻速。

none-disperesed drilling fluid *np.* 不分散钻井液

例：None-disperesed drilling fluid refers to the water-based drilling fluid treated by organic polymer aggregates with flocculation and coating.

不分散钻井液体系是指经过有机高分子集合物处理的水基钻井液，这些有机高分子具有絮凝及包被作用。

aerated mud *np.* 充气钻井液

例：The aerated mud can be used in negative drilling, and enlarge the life of drilling bits for its low density, and low sap pressure.

充气钻井液密度低，液相压力低，可用于负压钻井，延长钻头寿命。

gas-cut mud *np.* 气侵钻井液

例：Gas-cut muds can form hydrates in deepwater drilling operations, plugging BOP lines, risers and subsea wellheads, causing a well-control risk.

气侵钻井液会在深水钻井作业中形成水合物，堵塞防喷器管线、立管和海底井口，从而造成井控风险。

displacement fluid *np.* 顶替液 / 驱替液

例：Correctly calculating the appropriate volume of <u>displacement fluid</u> while taking account of well production are key factors in achieving accurate placement of fluids.

正确计算顶替液用量要把油井生产因素考虑进去，这对于准确注入钻井液至关重要。

water–in–oil emulsion *np.* 油包水乳化液

例：Oil-based mud are similar in composition to water-based mud, except the continuous phase is oil instead of water and water droplets are emulsified into the oil phase. This type of fluid is called a <u>water-in- oil emulsion</u>.

油基钻井液的成分与水基钻井液相似，不同之处在于连续相是油而不是水，水滴乳化到油相中，这种流体称为油包水乳液。

oil–in–water emulsion *np.* 水包油乳液

例：Any oil added to water-based mud is emulsified into the water phase and is maintained as small, discontinuous droplets. This type of fluid mixture is called an <u>oil-in-water emulsion</u>.

添加到水基钻井液中的油都会乳化到水相中，并保持为小而不连续的液滴。这种流体混合物称为水包油乳液。

AME/ thermal activated mud emulsion *np.* 热活化钻井液乳液

例：<u>Thermal activated mud emulsions</u> usually consist of surfactant-cosurfactant-solvent and water (or aqueous solution).

热活化钻井液乳液通常由表面活性剂、助表面活性剂、溶剂和水（或水溶液）组成。

slurry yield *np.* 水泥造浆量

例：<u>Slurry yield</u> is the volume of slurry obtained when one sack of cement is mixed with the desired amount of water and other additives.

一袋水泥与所需水量及其他添加剂混合即称水泥造浆量。

viscosifier *n.* 增黏剂

例：Adoption of <u>viscosifiers</u> is a key technique to ensure lightweight aggregate concrete with big fluidity to have good performances.

保证轻集料混凝土流动性强，具有良好性能，采用增黏剂是关键。

weighting material *np.* 加重材料

例：Mineral hardness, particle size and shape are the main parameters that affect abrasiveness of <u>weighting materials</u>.

矿物硬度、粒度和形状是影响加重材料耐磨性的主要参数。

temperature stability agent *np.* 高温稳定剂

例：<u>Temperature stability agents</u> are treatment agents that can perform its function continuously at high temperature.

高温稳定剂是一些在高温条件下能持续发挥其功能的处理剂。

surface active agent/surface-active chemical *np.* 表面活性剂

例：<u>Surface active agents</u> are widely used because of their properties such as significantly reducing interfacial tension, increasing solubility and increasing flow.

表面活性剂由于具有能显著降低界面张力、提高溶解度和提升流动性等特性而得以广泛应用。

surface tension *np.* 表面张力

例：To make a foam, as used for a drilling fluid, the liquid's <u>surface tension</u> must be lowered by adding a third component (a foamer).

为了制造泡沫，用于钻井液中，必须通过添加第三种成分（发泡剂）来降低液体的表面张力。

mud resistivity *np.* 钻井液电阻率

例：The influence of width, porosity, number of fracture and <u>mud resistivity</u> on result of forward simulation result was analyzed.

研究分析了裂缝宽度、裂缝孔隙度、裂缝条数和钻井液电阻率的变化对正演模拟结果的影响。

lubricant *n.* 润滑剂

例：Fatty acids are the raw materials used in the manufacture of many drilling-fluid additives, such as emulsifiers, oil-wetting agents and lubricants.
脂肪酸是制造许多钻井液添加剂的原材料，如乳化剂、润油剂和润滑剂。

foaming agent *np.* 发泡剂

例：Foaming agents are usually nonionic surfactants and contain polymeric materials.
发泡剂通常是非离子表面活性剂，内含聚合物材料。

filtrate reducer *np.* 降滤失剂

例：Filtrate reducer refers to the chemical agent that can reduce the fluid loss of drilling fluid.
降滤失剂是指能降低钻井液滤失量的化学剂。

emulsifier *n.* 乳化剂

例：Numerous types of emulsifiers will disperse oil into water muds, including sulfonated hydrocarbons, ethyoxylated nonylphenols, alkali-metal fatty-acid soaps, etc.
许多类型的乳化剂会将油分散到水泥浆中，包括磺化烃、乙氧基化壬基酚和碱金属脂肪酸皂等。

defoamer *n.* 消泡剂

例：Octyl alcohol, aluminum stearate, various glycols, silicones and sulfonated hydrocarbons are used as defoamers.
辛醇、硬脂酸铝、各种二醇、硅酮和磺化烃可作为消泡剂。

corrosion inhibitor *np.* 防腐蚀剂

例：Some sequestering agents, corrosion inhibitors or friction reducers can also form residues that may plug formation pores.
一些隔离剂、防腐蚀剂或减阻剂也会形成残留物，堵塞地层孔隙。

slurry stability *np.* 浆体稳定性

例：Slurry stability is the ability of a cement slurry to maintain homogeneity which is measured by the free-fluid test and the sedimentation test.

浆体稳定性是指水泥浆保持均匀性的能力，通常通过自由流体试验和沉降试验来测定。

plastic viscosity *np.* 塑性黏度

例：Formation solids contained in a mud system is generally considered to be detrimental to the drilling operation because they produce high plastic viscosity.

通常认为，钻井液系统中含岩屑会促使塑性黏度升高，不利于钻井作业。

apparent viscosity *np.* 表观黏度

例：As the shear rate increased, the apparent viscosity of the solution gradually decreased.

随着剪切速率的增大，溶液的表观黏度逐步下降。

funnel viscosity *np.* 漏斗黏度

例：Funnel viscosity is an important parameter that needs to be measured frequently during drilling.

在钻井过程中，漏斗黏度是需要经常测定的重要参数。

drilling fluid density *np.* 钻井液密度

例：Drilling fluid density refers to the mass of drilling fluid per unit volume.

钻井液密度是指单位体积钻井液的质量。

gelation characteristics *np.* 胶凝特性

例：In order to drill high-quality wells quickly, the gelation characteristics of drilling fluid must be adjusted according to different drilling conditions and requirements.

为了快速打出优质井，必须针对不同的钻井情况和要求调整好钻井液胶凝特性。

fluid circulating system *np.* 钻井液循环系统

例：The main components of a fluid circulating system for rotary drilling involve: the pump, hose and swivel, drill string, mud return line, and pits.

旋转钻井钻井液循环系统由泵、水龙带、水龙头、钻柱、钻井液返出管和钻井液池构成。

Section IV

Rotary Drilling
旋转钻井

drilling site selection *np.* 钻井选址

例：Because surface geology is an inexpensive and reliable method of <u>drilling site selection</u>, it was used extensively for many decades.

地面地质调查法是一种低成本、可靠的钻井选址方法，广泛应用了几十年。

right-of-way for access *np.* 场地使用权

例：Lease terms and agreements are thoroughly reviewed by legal experts for clear title and <u>right-of-way for access</u>.

法律专家对租赁条款和协议进行了彻底审查，以明确作业场地使用权。

well pad *np.* 井场

例：Horizontal drilling makes drilling multiple wells from a single well pad possible; enabling E&P companies to reduce costs and surface disturbance associated with <u>well pad</u> construction.

水平钻井可实现在一个井场钻多个分支井，使勘探开发公司降低井场建设成本、减少地面占用。

drilling operation *np.* 钻井作业

例：Normal <u>drilling operations</u> are, after all, what the rig and crew are mainly hired to do.

正常的钻井作业主要是钻机和钻井队工作。

cellar *n.* 方井

例：The rig is placed directly over the <u>cellar</u>, which provides extra space for drilling accessories that will be installed under the rig.

钻机直接放置在方井上方，方井可为钻机下方安装配件提供空间。

rigging up　*vp.* 安装钻机

例：The next step is rigging up, putting the rig together so that drilling can begin.

下一步是安装钻机，将钻机组装在一起，以便开始钻井。

lined with　*vp.* 内衬

例：This first part of the hole is large in diameter but fairly shallow in depth, lined with large-diameter pipe called conductor pipe.

开始的井眼直径较大，但深度相当浅，内衬一根大直径管道，称为导管。

rig floor　*np.* 钻台

例：The drilling line is reeved (threaded) over a crown block sheave and lowered down to the rig floor.

钻井绳穿过天车滑轮下落到钻台上。

rathole　*n.* 大鼠洞

例：Another hole (rathole) is dug off to the side of the cellar, also lined with pipe. The rathole serves as a place to temporarily store a piece of drilling equipment called the kelly.

方井旁另钻一个洞（大鼠洞），并内衬一根管子，用于临时存放一根方钻杆。

mousehole　*n.* 小鼠洞

例：The mousehole is a hole in the rig floor into which a length of pipe is placed.

小鼠洞是钻台上的一个洞，接钻柱前存放钻杆的地方。

making a connection　*np.* 接单根

例：What has been described is making a connection; it takes place each time the kelly is drilled down.

该过程就是接单根；每钻完一根方钻杆长度，就进行一次接单根作业。

disconnect/break out　*v.* 卸下/卸扣

例：A set of slips is a tapered device lined with strong, teethlike gripping elements that, when placed around drill pipe, keep pipe suspended in the hole when the kelly is disconnected.

卡瓦是锥形装置，内衬有坚固的牙齿状夹持元件。当方钻杆卸下时，将卡瓦卡在钻杆周围，使钻杆悬在井内。

make up *vp.* 连接 / 上扣

例：When pipe is <u>made up</u> (joined together), the pin is stabbed into the box and the connection tightened.

连接钻杆（把钻杆接合在一起）时，要将接头插入内螺纹，并将接口处拧紧。

trip out *vp.* 起钻

例：At this point, the drill string and bit are <u>tripped out</u> of the hole.

钻达这个层位时，就可以起出所有的钻柱和钻头。

monkey board *np.* 二层台

例：The derrickman, using a safety harness and climbing device, has climbed up into the mast or derrick to his position on the <u>monkey board</u>, a small working platform provided for him.

井架工系上安全带，利用攀爬装置爬上井架二层台。二层台是井架工在井架上的小型工作台。

fingerboard *n.* 指梁

例：As the top of the pipe reaches the derrickman's position, he guides it back into a rack called the <u>fingerboard</u>.

当钻杆顶部到达井架工的位置时，他将钻杆放到被称为指梁的架子上。

stand *n.* 立根

例：Three joints of pipe connected together constitute what is termed a <u>stand</u>, a thribble.

三根连接在一起的钻杆构成了一个立根，或称三节柱。

surface hole *np.* 表层

例：The drilling stops because this first part of the hole—the <u>surface hole</u>—is drilled only deep enough to get past soft, sticky formations, gravel beds, freshwater-bearing formations, and so forth that lie relatively near the surface.

第一部分井段（表层）只需钻过近地表较软黏性地层、砾石层、淡水层等，即可停钻。

running casing *np.* 下套管

例：Running casing into the hole is very similar to running drill pipe, except that the casing diameter is much larger and thus requires special elevators, tongs, and slips to fit it.

下套管与下钻杆非常相似，不同之处在于套管直径要大得多，因此需要特殊的吊卡、大钳和卡瓦来配合。

trip in *vp.* 下钻

例：When the drill string and bit are tripped out, a new bit, suitable for the type of formations being drilled, is made up, and the whole assembly of bit, drill collars, and drill pipe is tripped back in.

把钻柱和钻头起出，把一只适用于所钻地层类型的新钻头接在钻柱上，然后再把钻头、钻铤和钻杆下入井中。

round trip *np.* 起下钻

例：Several round trips may be necessary before drilling is once again brought to a halt.

可能还需要多次起下钻才能结束这次钻进。

annular space *np.* 环形空间

例：This drilling fluid is pumped from the surface down the inside of the rotating drill string, discharged through ports in the bit and returned to surface via the annular space between drill pipe and hole.

钻井液从地面泵送至旋转钻柱内，通过钻头孔眼排出，经钻柱与井壁之间的环形空间返回地面。

drill cuttings *np.* 钻屑

例：The drill cuttings are sampled at regular intervals and studied in the laboratory in much the same way as the surface samples.

定期对钻屑进行取样，并在实验室中以与地表样品大致相同的方式进行研究。

filter cake *np.* 滤饼

例：Solids added to a drilling fluid to bridge across the pore throat or fractures of an exposed rock thereby building a filter cake to prevent loss of whole mud or excessive filtrate.

将固体添加到钻井液中，以弥合暴露岩石的孔喉或裂缝，从而形成滤饼，防止整个钻井液或过多滤液流失。

fluid loss *np.* 滤失量

例：The packing efficiency decreases with the increase of gravel concentration in injected liquid, fluid loss to formation and gravel size.

携砂液滤失量、砾石浓度和砾石尺寸增加都会导致充填效率降低。

filtration control *np.* 滤失控制

例：A clay mineral that is composed principally of three-layer clays, and widely used as a mud additive for viscosity and filtration control.

黏土矿物，主要由三层黏土组成，且它广泛用作钻井液添加剂，用于黏度控制和滤失控制。

wall cake *np.* 井壁滤饼

例：Wall cake is formed after filtration loss on the well wall during drilling fluid circulation, which plays a role of stabilizing the well wall and ensuring smooth circulation and displacement of working fluid.

在循环过程中，钻井液滤失会在井壁上形成滤饼，滤饼可稳定井壁，保证钻井液循环和驱替顺利。

influx *n.* 井侵

例：In the case of an underbalanced kick, the driller must circulate the influx out and increase the density of the drilling fluid.

欠平衡井侵一旦发生，司钻必须将侵入的流体循环出去，并提高钻井液密度。

well kick *np.* 井涌

例：When a well kick occurs, it makes its presence known by certain things that happen in the circulating system.

当井涌发生时，循环系统就会有一定的显示。

kicking well *np.* 喷发前兆井

例：We need to take precautions against a kicking well.

我们要对有喷发前兆的油井采取预防措施。

blowout *n.* 井喷

例：Oil-well blowouts are wasteful, not only of time and money spent for control, but of pressure in the formation, which is needed to move the oil from the underground reservoir and raise it to the surface.

油井井喷是一种浪费，不仅浪费了用于控制的时间和金钱，还浪费了将石油从地下储层转移到地面所需的地层压力。

wild well *np.* 猛喷井

例：A well which blows out of control is known as a 'gusher' or a 'wild well'.

油井失控而发生井喷，称为"喷射井"或"猛喷井"。

gas cut *np.* 气侵

例：A gas cut is inferred only if the mud returning to the surface is significantly less dense than it should be.

只有当返回地面的钻井液密度明显低于应有密度时，才能推断出气侵。

water influx *np.* 水侵；含水率

例：According to solution of optimal model, geologic reserves and water influx value can be obtained directly, and water range and water cut coefficient can also be determined.

通过优化模型求解直接获得地质储量及水侵量的大小，同时还可以确定水域大小和含水率。

choke manifold *np.* 阻流管汇

例：The choke manifold plays a vital part in protecting the well in that its use controls the mud column in a kicking well.

阻流管汇在保护油井方面起着至关重要的作用，因其控制了井涌油井中的钻井液柱。

killing the well *np.* 压井

例：The procedure for killing the well and stabilizing the mud column has been described earlier.

压井和稳定钻井液柱的程序已在前面描述。

primary well control *np.* 一级井控

例：Primary well control is the drilling process in which the downhole pressure is slightly higher than the formation pressure by using appropriate drilling fluids and technical measures to control formation pressure so that no formation fluids enter the well.

一级井控依靠适当的钻井液和技术措施来控制地层压力，使井底压力稍大于地层压力，阻止地层流体进入井内。

secondary well control *np.* 二级井控

例：The ram BOP is the most important equipment for secondary well control, quick to shut in the well, and of a high pressure rating.

闸板防喷器关井速度快、耐高压，是二级井控最重要的设备。

tertiary well control *np.* 三级井控

例：Tertiary well control refers to the process of restoring control of a well after it has blown out of control, using appropriate technology and equipment.

三级井控是指井喷失控后，使用适当的技术和设备抢险，恢复对井的控制。

lost circulation *np.* 井漏

例：If the pressure in the formation is lower than the pressure exerted by the mud column, then normal operations can continue unless the formation is so porous or fissured that it allows the mud to pass into the reservoir, which is "lost circulation".

如果地层压力低于钻井液柱压力，则正常钻井作业可以继续，除非地层孔隙性很大或有裂缝使钻井液流入储层，这种情况就是"井漏"。

kick off *vp.* 造斜

例：Only the first hole drilled into the reservoir may be vertical; every subsequent well may be drilled vertically to a certain depth, then kicked off (deflected) directionally.

只有第一口井垂直钻入储层；随后的每口井可垂直钻至一定深度，然后定向造斜（偏转）。

kick-off point *np.* 造斜点

例：At the kick-off point, the hole which is drilled through the casing is referred to as the "window".

在初始造斜点穿过套管而钻的孔称为"窗口"。

whipstock *n.* 造斜器

例：Another directional drilling tool is the old, reliable whipstock.

另一种定向钻井工具是旧的、可靠的造斜器。

sidetrack inside casing *np.* 套管内侧钻

例：Sidetrack inside casing is a very effective way to exploit the residual oil and gas with low cost.

套管内侧钻是低成本开发剩余油气的有效途径。

hang sidetrack drilling technology *np.* 悬空侧钻技术

例：During drilling operation, the track in horizontal interval is controlled and the hang sidetrack drilling technology is successfully used.

钻井作业中精确控制了水平段轨迹并应用了悬空侧钻技术。

slant hole *np.* 斜井 / 斜孔

例：The slant hole tray was applied to the technical innovation of vinyl acetate distillation column.

采用复合斜孔塔板对醋酸乙烯精馏塔进行了技术改造。

well trajectory *np.* 井眼轨迹

例：Well trajectory control is one of the key technologies of high-angle well drilling, and the results of the control directly affect the well friction, torque, safety and construction of all wells.

井眼轨迹控制是大斜度井钻井的关键技术之一，其结果直接影响该井摩阻、扭矩的大小、全井的安全及施工等。

hole curvature *np.* 井眼曲率

例：The hole curvature can be controlled through controlling the magnetic filed intensity and the velocity of electron.

控制电子束运动速度和磁场强度，即可控制井眼曲率。

horizontal displacement *np.* 水平位移

例：Deep horizontal displacement scarcely increased after unloading and reloading.

卸载再加载后，深层水平位移基本不增加。

vertical section displacement *np.* 视平移 / 垂直截面位移

例：Vertical section displacement is an important parameter in drawing vertical projection map.

视平移是绘制井身垂直投影图的重要参数。

deviation correction *np.* 纠斜

例：Combination of flexible deviation correction and prevention drilling tools can control the increase of deviation of the drilling well effectively.

柔性纠斜防斜钻具组合能有效地控制井斜增加。

azimuth angle *np.* 方位角

例：The measurement of the inclination angle and the azimuth angle of a well in the drilling process is important.

钻探过程中倾斜角和方位角的测量至关重要。

azimuth change rate *np.* 方位变化率

例：Azimuth change rate is the change of azimuth per unit length.

方位变化率是指单位长度内的方位角变化情况。

inclination azimuth *np.* 井斜方位

例：The tool consists of a Combined Inclination Azimuth Multi-function Logging Tool and a MSC 36-Arm Caliper Imaging Logging Tool.

该仪器由组合式井斜方位多功能测井仪和 MSC-36 多臂井径成像测井仪组合而成。

rate of inclination change *np.* 井斜变化率

例：The rate of inclination change is equal to the curvature of the well on the vertical profile.

井斜变化率等于井身在垂直剖面上的曲率。

line slope *np.* 造斜率

例：The influencing regular pattern of depths of deflection point, line slope and casing dimension on horizontal force of directional wellhead is analyzed by case calculation.

通过算例，分析了造斜点井深、造斜率以及套管尺寸对井口水平力的影响规律。

hold/maintain angle *vp.* 稳斜

例：Maintaining angle can rapidly increase the horizontal displacement of bottom hole.

稳斜钻进可以迅速增大井底水平位移。

angle drop-off *np.* 降斜

例：Shallow cone, short profile and short gage design to satisfy angle building and angle drop-off requirements.

浅内锥、短剖面和小直径设计，可满足造斜和降斜的需要。

drill stem test *np.* 地层测试

例：In a drill stem test, the testing assembly comprising a "packer", a "tester" and a recording pressure gauge are run into the well on empty, or partially empty, drill pipe, and positioned as near as practicable above the formation to be tested.

地层测试时，用中空或部分中空钻杆下入"封隔器""测试仪"和压力记录表等测试组件，并尽可能靠近目标测试地层上方。

hole cleaning *np.* 井眼清洁

例：Hole cleaning detecting is efficient by monitoring annular pressure change.

监测环空压力变化可有效检测井眼清洁程度。

orientation *n.* 定向

例：The drill string rotates several times during running, so there must be some technical measures to ensure that the deflecting tool points at the bottom of the hole to the predetermined azimuth of the device. This process is called <u>orientation</u>.

下钻过程中钻柱要多次旋转，所以要有一定的措施保证造斜工具在井底指向预定的装置方位角，此过程就是定向。

orientational coring *np.* 定向取心

例：This technology of <u>orientational coring</u> can not only obtain conventional core information, but also elements of attitude of formation and fractures (dip angle).

岩心定向取心技术不仅能获得常规岩心资料，还能获得地层和裂缝的产状要素（倾角、倾向）。

measurement while drill (MWD) *np.* 随钻测量

例：<u>Measurement while drilling</u> is one of key technologies for monitoring and controlling well trajectory.

随钻测量是井眼轨迹监测与控制中的一项关键技术。

plugging and abandoning a well *np.* 堵塞和弃井

例：The cost of <u>plugging and abandoning a well</u> may only be a few thousand dollars.

堵塞和弃井的成本可能只有几千美元。

Section V

Well Cementation and Completion
固井与完井

cement job *np.* 固井作业

例：After the cement hardens, tests may be run to ensure a good <u>cement job</u>.
水泥固结后要进行检测，以确保良好的固井作业质量。

cementing unit *np.* 固井装置

例：The treated mud is pumped to a <u>cementing unit</u>, where the slag is added.
处理后的钻井泵入固井装置，加入矿渣。

casing program *np.* 下套管工作

例：Well-heads and <u>casing programs</u> vary considerably in detail.
井口安装和下套管在细节上差别很大。

conductor casing *np.* 导管

例：<u>Conductor casing</u> (outermost pipe) prevents collapse of loose soil near the surface and enables circulation of drilling fluids during drilling.
导管（最外层的套管）可防止近地表松散土壤坍塌，并在钻井过程中循环钻井液。

surface casing *np.* 表层套管

例：<u>Surface casing</u> protects freshwater zones near the surface from contamination from leaching fluids during drilling and gas during production.
表层套管可以阻止钻井液和采出气污染近地表的淡水层。

intermediate casing *np.* 中间套管

例：<u>Intermediate casing</u> is used in certain sections to protect the well from various hazardous subsurface conditions such as abnormal pressure zones and salt water deposits.

中间套管用于某些层段，使油气井免于压力异常地层和盐水层等各种地下危险情况。

production casing *np.* 生产套管

例：<u>Production casing</u> (innermost pipe) provides a conduit for the production tubing, which is ultimately inserted inside the production casing to carry the extracted gas to the surface.

生产套管（最里层套管）为油管导管，油管最后下入生产套管内，将采出气输送到地面。

casing accessory *np.* 套管配件

例：Other <u>casing accessories</u> include a guide shoe and a float collar.

其他套管配件包括套管鞋和浮箍。

guide shoe *np.* 套管鞋

例：A <u>guide shoe</u> is one of casing accessories.

套管鞋是套管配件之一。

float shoe *np.* 浮鞋

例：The <u>float shoe</u> also guides the casing toward the center of the hole to minimize hitting rock ledges or washouts as the casing is run into the wellbore.

浮鞋将套管导向井眼中心，以减少套管下入井筒时对岩壁的冲击。

float collar *np.* 浮箍

例：A <u>float collar</u> is designed to serve as a receptacle for cement plugs and to keep drilling mud in the hole from entering the casing.

浮箍设计成水泥塞的容器，用于防止井内钻井液进入套管。

centralizer *n.* 扶正器

例：<u>Centralizers</u> are attached to the casing and, since they have a bowed-spring arrangement, keep the casing centered in the hole after it's lowered in.

由于扶正器为弓形弹簧结构，因此把它安放到套管上，就可以保持套管在井眼中心。

scratcher *n.* 刮泥器

例：Also, devices called centralizers and scratchers are often installed on the outside of the casing before it is lowered into the hole.
此外，在套管下入井眼之前，通常将扶正器和刮泥器安装在套管外部。

casing property *np.* 套管特性

例：Tests in kind simulating the deep and ultra-deep well conditions are effective ways to testify the casing properties.
模拟深井和超深井条件试验是验证套管特性的有效途径。

casing strength *np.* 套管强度

例：The main reason of casing failure in oilfield is that the casing strength is insufficient to resist the immense external force in salt interval.
油田套管损坏的主要原因是现有套管强度不够，无法抵抗盐层巨大外力。

collapse strength *np.* 抗压强度

例：The collapse strength of casing is an active research field in the world.
套管抗压强度是目前国际上较为热门的研究领域。

bond strength *np.* 胶结强度

例：Bond strength refers to how well cement is bonded to casing or formation.
胶结强度指水泥与套管或地层之间的胶结程度。

safety factor *np.* 安全系数

例：Great care is taken in the selection of each casing string so that throughout its length it will possess a safety factor adequate to withstand the stresses and pressures which could occur in the well.
每节套管筛选都非常谨慎，以保证其足以承受井内可能产生的应力和压力。

multiple string cementing *np.* 多管注水泥法

例：The multiple string cementing is a method in which multiple tubing is used as casing in a well for more economic exploitation in a well with multiple oil layers.
多管注水泥法指在多油层井内，为更经济地开采，在一个井眼内下入多根油管作套管。

primary cementing *np.* 初次注水泥

例：Conventional <u>primary cementing</u> is to inject cement slurry through the casing, and by the upper and lower rubber plug isolation displacement, so that the cement slurry back out of the pipe.

初次注水泥一般指通过套管注入水泥浆，并由上、下胶塞隔离驱替，使水泥浆返出管外的固井方法。

buffer fluid *np.* 缓冲液

例：The mobility of <u>buffer fluids</u> assures a viscosity transition and mobility control.

缓冲液具有流动性，可以控制流体流动和黏度变化。

spacer *n.* 隔离液

例：The <u>spacer</u> is prepared with specific fluid characteristics, such as viscosity and density and is engineered to displace the drilling fluid while enabling placement of a complete cement sheath.

该隔离液具有特殊的流体特性，如黏度和密度，在水泥环完整时可用于驱替钻井液。

lead slurry *np.* 前导浆

例：<u>Lead slurries</u> refer to a section of cement with a high water cement ratio injected before cementing in order to improve displacement efficiency.

为提高驱替效率，固井前所注的一段高水灰比的水泥就是前导浆。

tail slurry *np.* 尾浆

例：<u>Tail slurries</u> refer to the cement slurry injected behind the lead slurry to isolate the target zone from the casing shoe.

尾浆指在注入前导浆后，用来封隔目的层和套管鞋而注入的水泥浆。

tail fluid *np.* 后置液（压塞液）/尾液

例：<u>Tail fluids</u> refer to a specially prepared displacement fluid when pressed into the glue plug after cementing.

后置液指固井后，压入上胶塞时用的特殊置换液。

cementing plug *np.* 注水泥塞

例：Two types of cementing plugs are typically used on a cementing operation.

固井作业通常使用两种注水泥塞。

annular velocity *np.* 环空流速

例：The annular velocity is commonly expressed in units of feet per minute.

环空流速通常用英尺/分钟来表示。

back flush *np.* 反冲洗

例：The process of back flush is the last step of deep bed filter.

反冲洗是深层过滤的最后环节。

bridge plug *np.* 桥塞

例：The well designer may choose to set bridge plugs in conjunction with cement slurries to ensure that higher density cement does not fall in the wellbore.

设计人员可以选择与水泥浆配合设置桥塞，以确保密度更高的水泥浆不会落在井中。

batch mixing *np.* 二次混浆

例：Batch mixing is a method that cement slurry is stirred twice through a certain container in advance and then injected into the well.

二次混浆指将水泥浆提前通过特定容器进行二次搅拌，然后注入井内。

waiting on cement (WOC) *np.* 候凝

例：After the cement is run, a waiting time is allotted to allow the slurry to harden. This period of time is referred to as waiting on cement or simply WOC.

完成注水泥后要等水泥浆硬化，这段时间称为候凝时间或WOC。

cement evaluation *np.* 水泥评估

例：After casing is cemented in, a cement evaluation is conducted.

套管注入水泥后要进行注水泥评估。

pay zone /the formation of interest *np.* 产层/目的层

例：Once again several bits will be dulled and several round trips will be made, but before long the formation of interest (the pay zone, the oil sand, or the formation that is supposed to contain hydrocarbons) is penetrated by the hole.

经几次起下钻更换钻头，不久就能钻达目的层（产层、油砂或可能含有碳氢化合物的地层）。

completion interval *np.* 完井层段

例：It should be isolated with a plug or packer to prevent damage to the completion interval while the other lateral is being drilled.

应使用桥塞或封隔器加以封隔，以防止在钻另一个分支井时损坏完井层段。

dual completion *np.* 双层完井

例：In some simple dual completions, the second or upper zone is produced up the tubing-casing annulus.

在一些简单的双层完井中，第二层或上层沿油管—套管环空进行。

lower completion *np.* 生产完井

例：Completion essentially involves preparing the bottom of the well hole to specifications designed to maximize gas production, installing production tubing and related downhole tools, perforating and stimulating the well (collectively called "lower completion") and installation of wellhead equipment.

完井基本上是按设计规格进行井底作业以实现产气量最大化，安装油管和相关井下设备，射孔并增产（总称"生产完井"），同时安装井口设备。

recompletion *n.* 二次完井

例：The integrated reservoir model provides the basis for geologically targeting potential infill and step-out drilling locations, recompletions, and field management strategies.

一体化油藏模型为地质上确定加密井和探边井的井位、二次完井以及制定油田管理策略提供了基础。

stimulation *n.* 增产

例：Stimulation is a treatment performed on hydrocarbon-bearing formations to improve the flow of gas from the formation to the wellbore.

增产是对含烃地层进行处理，以提高气体由地层到井筒的流速。

multiple completion *np.* 多层 / 多次完井

例：The wellhead and surface flow-control facilities required for multiple completions can be complex and costly.

多层 / 多次完井所需的井口和地面流量控制设施既复杂又昂贵。

screen liner *np.* 筛管 / 带眼衬管

例：The designs of screen liners for sand control and packer are very important in the small diameter screen liner technique.

在小直径筛管防砂技术中，防砂筛管和封隔器的设计至关重要。

casing head *np.* 套管头

例：The casing head, together with the casing hanger later, is known as the "permanent" well-head.

套管头以及后面装的套管悬挂器统称为"永久"井口装置。

tubing *n.* 油管

例：Small-diameter pipe called tubing is placed in the well to serve as a way for the oil or gas to flow to the surface.

将称为油管的小直径管置于井中，使油气从中流到地面。

packer *n.* 封隔器

例：The packer goes on the outside of the tubing and is placed at a depth just above the producing zone.

封隔器位于油管外侧，放置在产层正上方。

cased-hole perforated completion *np.* 套管射孔完井

例：Cased-hole perforated completions are commonly used in shale gas development.

页岩气开发一般使用套管射孔完井法。

open-hole completion *np.* 裸眼完井

例：Advances in completion technology have led to the emergence of open-hole multi-stage fracturing systems for horizontal wells and open-hole completions have also been utilized in many shale plays.

完井技术的进步促使了水平井裸眼多段压裂技术的出现，裸眼完井技术也已广泛应用于页岩气开发。

perforation n. 射孔

例：Since the pay zone is sealed off by the production string and cement, <u>perforations</u> must be made in order for the oil or gas to flow into the wellbore.

由于产层由生产套管和水泥封隔，因此必须进行射孔，以使石油或天然气流入井筒。

shaped–charge explosive np. 聚能射孔弹

例：The most common method of perforating incorporates <u>shaped-charge explosives</u> (similar to those used in armorpiercing shells).

最常见的射孔方法是使用聚能射孔弹（类似于穿甲弹）。

overbalanced perforating np. 正压/过平衡射孔

例：Where <u>overbalanced perforating</u> techniques are used, it may be necessary to acidize the crushed zone to achieve maximum productivity from the perforated interval.

正压射孔可能需要酸化破碎带，以使射孔段实现最大产能。

underbalanced perforating np. 负压/欠平衡射孔

例：Measures to reduce the effect of the crushed zone include <u>underbalanced perforating</u> in which the crushed zone and perforating debris are flushed from the perforating tunnel by the reservoir fluid as soon as the perforation is created.

减少压实带影响可用负压射孔技术，即射孔后，储层流体从孔道冲出破碎带和射孔碎屑。

hydraulic fracturing np. 水力压裂

例：<u>Hydraulic fracturing</u> involves isolating sections of the well in the producing zone and then pumping high volumes of specially engineered fracturing fluids at high pressure down the wellbore and out into the shale formation.

水力压裂作业是先把产层井筒分段，然后沿井筒高压泵入大量特制压裂液，压裂液从井筒出去进入页岩层。

single-stage process *np.* 单段压裂

例：Fracturing for coalbed methane production is frequently a single-stage process, *i.e.* one fracturing job per well, rather than multi-stage.

煤层气开采通常采用单段压裂，即每口井只进行一次压裂，而不是多段压裂。

multi-stage fracturing *np.* 多级压裂

例：Multi-stage fracturing tools have increased the speed and effectiveness with which long horizontal laterals can be fractured.

多级压裂工具的出现也使长距离水平段压裂的效率有所提高。

fracturing fluid *np.* 压裂液

例：Because productive coal seams are often at shallower depths than tight or shale gas deposits, there is also a greater risk that fracturing fluids might find their way into an aquifer directly or via a fracture system.

由于煤层气产层往往比致密气和页岩气藏浅，因而压裂液直接或间接通过裂隙进入含水层的风险也更大。

slickwater fracturing *np.* 滑溜水压裂

例：Two key advantages of foam fracturing over high-pressure slickwater fracturing are reduction of formation damage that could lead to blocked fractures and substantial reduction of the volume of water that is needed for the fracturing.

较之高压滑溜水压裂，泡沫压裂主要具有两大优势：一是减少了地层损害，避免造成裂缝堵塞；二是大大减少了压裂用水量。

foam fracturing *np.* 泡沫压裂

例：Foam fracturing technology finds use in lower permeability and lower pressure shales where particular consideration must be made to prevent induced formation damage and blockage from clay migration.

泡沫压裂技术一般用于渗透率和压力较低的页岩地层，对于这样的地层尤其要注意防止诱发性损害，以及黏土运移导致的堵塞。

nitrogen or carbon dioxide fracturing *np.* 氮气或二氧化碳气体压裂

例：Another fracturing technique suitable for underpressurized, water-sensitive shales is nitrogen or carbon dioxide fracturing.

另一种适合用于低压、水敏性页岩气藏的压裂技术是氮气或二氧化碳气体压裂。

non-aqueous fracturing fluid *np.* 无水压裂液

例：This involves the use of a non-aqueous fracturing fluid, such as 100 percent nitrogen (or CO_2) as the carrier for specially engineered light-weight proppants.

该方法需使用无水压裂液，如用100%的氮气（或二氧化碳）携带特制轻质支撑剂。

proppant *n.* 支撑剂

例：Proppants, such as sand with a specific grain size, mixed in the fracturing fluid settle in the fractures and essentially prop the fractures open to allow continual gas flow.

将大小合适的砂子等支撑剂混入压裂液，支撑剂进入裂缝并保持裂缝张开，气体即可不断流动。

flow-back fluid *np.* 返排流体

例：The flow-back fluids recovered from the well are pumped to lined containment pits or tanks for treatment or off-site disposal.

井中返排流体泵入防渗漏的储存池或储存罐中，供后续处理或排放。

frac tree *np.* 压裂树

例：Before fracturing operations, a wellhead, or "frac tree" designed specifically for hydraulic fracturing is installed.

在进行压裂作业之前，要先安装井口设备，即为水力压裂专门设计的"压裂树"。

flowback equipment *np.* 返排设备

例：Additionally, flowback equipment including pipes, manifolds, and a gas-water separator, are installed at the surface and the system is pressure tested.

此外，还要在地表安装管汇和气水分离器等返排设备，并对系统进行压力测试。

high-pressure wellhead *np.* 高压井口装置

例：After fracturing, the frac tree is replaced with a <u>high-pressure wellhead</u> that remains in place to monitor and regulate the flow of gas and other fluids out of the well during gas production.

压裂结束后，用高压井口装置替换压裂树，监测并调节流出井口的天然气及其他流体。

Christmas tree *np.* 采油树

例：When casing is set, cemented, and perforated and when the tubing string is run, then a collection of valves called a <u>Christmas tree</u> is installed on the surface at the top of the casing.

完成了下套管、固井和射孔以及下油管等工作后，一组阀门（称为采油树）被安装在套管顶部。

Section VI

Well Logging and Remedial Work
测井与修井

well logging *np.* 测井

例：<u>Well logging</u> is one of the most common and most important methods for stratigraphic comparison, for tectonic analysis, and for pay evaluation.

测井是地层对比、构造分析和产层评价最常用和最重要的方法之一。

formation evaluation *np.* 地层评价

例：<u>Formation evaluation</u> involves measuring and analyzing specific petrophysical properties of the rock in the immediate vicinity of a wellbore to determine a formation's boundaries, volume of hydrocarbons, and ability to produce fluids to the surface.

地层评价是通过测量和分析井眼周围岩石的物理性质，确定地层边界、油气储量和可采流体。

geophysical log *np.* 地球物理测井数据

例：The well log may be based either on visual inspection of samples brought to the surface, <u>geological logs</u>, or on physical measurements made by instruments lowered into the hole, geophysical logs.

测井曲线可以基于对带到地表的地下岩层样品目测所得的地质录井数据，也可基于下入井中测井仪进行物理测量所得的地球物理测井数据。

geological log *np.* 地质录井数据

例：<u>Geological logs</u> use data collected at the surface rather than by downhole instruments.

地质录井数据来源于地表，而不是井下仪器。

mud logging *np.* 钻井液录井

例：In <u>mud logging</u>, logs are made by examining the bits of rock circulated to the surface by the drilling mud in rotary drilling.

在钻井液录井作业中，录井曲线图是通过检测旋转钻井中钻井液循环上来的岩屑做出的。

mud log *np.* 钻井液录井图

例：In addition to lithology, <u>mud logs</u> also typically include real-time drilling parameters such as rate of penetration, temperature of the drilling fluid, and chlorides.

除了岩性，钻井液录井图通常还包括实时钻井参数，如钻进速度、钻井液温度和氯化物等。

logging car *np.* 测井车

例：The <u>logging car</u> usually is a heavy truck which has to carry the cable and the cable reel, the logging cabin and similar equipment.

测井车通常是重型卡车，载有电缆、绞盘、测井箱体等设备。

borehole log/well log *np.* 测井曲线

例：Despite the importance of these data (which may be supplemented by cores of rock cut in the borehole), the basic geological data for subsurface interpretation is obtained in the form of <u>borehole logs</u>.

尽管这些数据很重要（可由钻孔中的岩心补充），但解释地下的基本地质数据还是要从测井曲线中获得。

wireline logging *np.* 电缆测井

例：<u>Wireline logging</u> is conducted by pushing instruments through the wellbore after it is drilled to evaluate well integrity, perform cement evaluations in the well casing process.

电缆测井是在钻井完成后沿井筒下入测量仪器的，可用于评价油气井完整性，以及油气井下套管过程中注水泥情况。

nuclear magnetic resonance (NMR) logging tool *np.* 核磁共振测井仪

例：<u>Nuclear magnetic resonance (NMR) logging tools</u> are an innovative new technology being used to measure rock porosity, estimate permeability from pore size distribution and identify pore fluids (water, oil and gas).

核磁共振（NMR）测井仪器是一种新技术，用于测量岩石孔隙度，根据孔隙大小分布计算渗透率，并识别孔隙内流体（如水、石油和天然气）。

electrical log *np.* 电测井

例：There are many different logs that can be obtained from different sondes, but the basic <u>geological log</u> is the "electrical log".

利用不同的探测仪可能得到许多不同的测井曲线，但基本的地质测井是"电测井"。

acoustic log/sonic log *np.* 声波测井

例：<u>Acoustic logs</u> either measure the velocity of sound within the formation, or the attenuation of sound.

声波测井测量岩层中的声波速度或声波的衰减。

pulse of ultrasonic sound *np.* 超声波脉冲

例：The probe contains an emitter, which produces <u>pulses of ultrasonic sound</u>, and two or more receivers.

探头包含一个产生超声波脉冲的发射器和两个或多个接收器。

nuclear Logging *np.* 核测井

例：Sealed neutron source and gamma source used in <u>nuclear logging</u> have strong radioactivity which is harmful to operators.

核测井使用的密封中子源和伽马源具有较强的放射性，对作业人员有一定的辐射损伤。

radioactivity log *np.* 放射性测井

例：The commonest of "<u>radioactivity logs</u>" is the gamma ray log, which records the rate of spontaneous emissions of gamma rays from the sedimentary rock.

最常见的"放射性测井"是伽马射线测井，记录沉积岩中伽马射线的自发发射率。

gamma ray log *np.* 伽马射线测井

例：In general, the radioactivity logged by the gamma ray log shows the shale contents of the rock, and that is a very simple but useful log.

一般来说，伽马射线测井记录的放射性可以显示岩石中的页岩含量，这是一种非常简单但有用的测井方法。

radioactive isotope *np.* 放射性同位素

例：This natural gamma activity mostly dates from a radioactive isotope of potassium, called potassium.

这种自然伽马放射主要源于一种被称为钾的放射性同位素。

density log *np.* 密度测井

例：The density log has been developed to determine the porosity of rock, the compaction of shales and similar details by nuclear methods.

密度测井已发展为基于核方法来确定岩石孔隙度、页岩压实度等情况。

logging while drilling (LWD) *np.* 随钻测井

例：In the logging while drilling (LWD) process, instruments are attached to the drill string and take measurements while the well is being drilled.

在随钻测井中，测井仪器安装在钻柱上，随钻井进程进行测量。

combination logging *np.* 组合测井

例：Combination logging tools consist of more than one sonde and cartridge, so that more than one measurement can be recorded on a single trip into the wellbore.

组合测井工具由多个探测仪和盘根组成，因此一次下钻可以记录多个测量值。

induction log *np.* 感应测井

例：This conductivity will cause an induction log to read too low a resistivity, by an amount that depends on its depth of investigation and the radius of the annulus.

这种导电性会导致感应测井电阻率读数过低，其具体数值取决于探测深度和环空半径。

log heading *np.* 测井曲线图头

例：The well name, location, casing information, and logging equipment data are found on the log heading.

井的名称、位置、套管信息和测井设备等数据都可在测井曲线图头中找到。

logging cable　*np.*　测井电缆

例：The power to the probe is transmitted into the borehole by means of the logging cable itself.

井中探测器的动力是通过测井电缆传输的。

sonde　*n.*　探头

例：To perform a logging operation, the measuring instrument, often called a sonde, is lowered into the borehole on the end of an insulated electrical cable.

测井作业通常需要将称为探头的测量仪器下入井筒，并置于绝缘电缆的末端。

transmitting coil　*np.*　发射线圈

例：In a typical implementation, the two arrays share the same transmitting coil but have different receivers.

典型的实例是，两种线圈系共用发射线圈，但是有各自的接收线圈。

well workover　*np.*　修井

例：Along with the development of oil and gas production, well workover has become a indispensable work.

随着油气开发的不断深入，修井已成为一项必不可少的工作。

open-hole　*np.*　裸眼井

例：Open-hole fishing nearly always takes place with mud in the hole, so the hazard of stuck pipe by differential pressure must be considered.

裸眼井打捞几乎井中都有钻井液，因此必须考虑压差造成的卡钻危险。

cased-hole　*np.*　套管井

例：Open-hole and cased-hole fishing jobs involve somewhat similar tools and techniques, but the problems and hazards differ.

裸眼井和套管井打捞作业的工具和技术基本相同，但问题和危险不同。

service well-head *np.* 修井井口装置

例：The <u>service well-head</u> is a combination of two blowout preventers (BOPs), a sleeve type and a double-ram type.

修井井口装置由两个防喷器组成，一个是套筒式防喷器，另一个是双闸板式防喷器。

fishing *n.* 打捞

例：The recovery operation for stuck pipe, packers, and loose equipment in a well is generally termed "<u>fishing</u>".

打捞井内被卡的钻杆、封隔器和其他井下落物的作业通常称为"打捞"。

fish *n.* 落鱼

例：A "<u>fish</u>" is part of a string of pipe or any other sizable piece of metal that might be loose in a well.

"落鱼"是落入井中的部分钻杆或其他任何有一定尺寸的金属材料。

junk basket *np.* 打捞篮

例：There are many types of fishing tools. For example, there is a type of fishing tool called a "<u>junk basket</u>", and there is another type called a "spear".

打捞工具有很多种。例如，有一种打捞工具称为"打捞篮"，还有另一种称为"矛类打捞工具"。

Part III

Oil and Gas Exploitation
油气开发

Section I

Fundamentals of Petroleum Production
油气开采基本概念

well pattern *np.* 井网

例：A suitable well pattern is chosen and steam is injected into a number of wells while the oil is produced from adjacent wells.

选择合适井网，将蒸汽注入多口井，石油从相邻的油井中采出。

well density *np.* 井网密度

例：The rational well spacing under certain perforating level and the rational open thickness under certain well density can be determined by this equation.

用此关系式可确定该类油藏在给定打开程度下的合理井距及给定井网密度下油井的合理打开厚度。

the degree of oil-affinity *np.* 亲油化度

例：The effect of modification is characterized by the degree of oil-affinity and the water-sorption rate.

用亲油化度和吸水率表征改性效果。

reserve-production ratio *np.* 储采比

例：The remaining reserves and reserve-production ratio of oil and gas fields can be predicted based on Weibull Model.

基于威布尔模型可以预测油气田剩余储量和储采比。

initial formation pressure *np.* 原始地层压力

例：The initial formation pressure should be retained in deformed media reservoir development in order to obtain a higher recovery.

保持原始地层压力开发变形介质油藏，才能得到较高采收率。

reservoir pressure *np.* 储层压力

例：Oilfield water injection is the main measure to maintain the <u>reservoir pressure</u>.

注水是油田保持储层压力的主要措施之一。

interference between layers *np.* 层间干扰

例：Zone replacement is studied aiming at the problems of high water cut and serious <u>interference between layers</u>.

针对油田开发中后期存在的综合含水高、层间干扰问题突出等，研究了换层采油技术。

sucker rod pumping *np.* 有杆抽油

例：<u>Sucker rod pumping</u> systems are the oldest and most widely used type of artificial lifting system for oil wells.

有杆抽油系统是油井生成中应用最久和最广泛的人工举升系统。

capillary pressure *np.* 毛细管压力

例：The motivating force of primary migration is the coalition of buoyancy and <u>capillary pressure</u>.

油气运移的动力来自浮力和毛细管压力的共同作用。

water sensitivity *np.* 水敏

例：Since <u>water sensitivity</u> was discovered, it has been studied for more than 70 years in several fields such as petroleum industry, geotechnical engineering and landfill remediation

水敏性自发现以来，已在石油工业、岩土工程和垃圾填埋场修复等多个领域进行了70多年的研究。

alkali sensitivity *np.* 碱敏

例：<u>Alkali sensitivity</u> damage occurred at both inlet and outlet sides of the core, suggesting that it would impact both proximal and distal zones from an injection well.

岩心的入口侧和出口侧均出现碱敏损伤，可能会影响注入井的近端和远端区域。

injection well *np.* 注水井

例：Onshore waterflood generally involves the drilling of several <u>injection wells</u> and the conversion of several existing wells into injection wells.

陆上水驱作业通常要钻几口注水井，同时还要将几口钻好的井转为注水井。

productivity index (PI) *np.* 采油指数

例：PI, the abbreviation for <u>productivity index</u>, is the ratio of recovered resources to total recoverable resources within the extraction area.

PI 是采油指数的缩写，指采区内采出的油气占可采油气资源总量的比值。

host facility *np.* 中心处理设施

例：In offshore production, oil and gas are extracted from the wells and brought to the surface to a <u>host facility</u> above the ocean surface.

海上采油时，油气从油井中抽出来后输送到海面上方的中心处理设施。

Section II

Conventional Oil and Gas Recovery
常规油气开发

conventional reservoir *np.* 常规油气藏

例：While hydraulic fracturing is already used on occasions to stimulate <u>conventional reservoirs</u>, tight gas and shale gas developments almost always require the use of this technique.

常规油气藏增产有时也用水力压裂技术，但致密气和页岩气开发几乎离不开这种技术。

producing formation *np.* 生产层

例：The production casing completely seals off the <u>producing formation</u> from water aquifers.

生产套管将产层与含水层完全隔开。

tubing *np.* 油管

例：The reservoir is only a part of a larger system that includes the reservoir, wellbore, <u>tubing</u>, artificial lift equipment, surface control devices, separators, treaters, tanks, and metering devices.

油气生产大系统包括油藏、井筒、油管、人工举升设备、地面控制装置、分离器、节流器、油罐以及测量装置，油藏只是其中一部分。

auxiliary tubing string *np.* 辅助油管

例：The safety joint allows for the parting of an <u>auxiliary tubing string</u> beneath a multiple string packer when the packer is being retrieved.

当回收封隔器时，安全接头可以帮助分离多管柱封隔器下方的辅助油管。

tubing head *np.* 油管头

例：The tubing head supports the tubing string, seals off the space between the casing and the inside of tubing and provides connections at the surface to control the flowing liquid or gas.

油管头支撑油管柱，封住油管和套管之间的环形空间，在地面就可以控制液体或气体的流动。

tubing anchor *np.* 油管锚

例：The tubing anchor is essentially a packer without the sealing element and is designed to prevent tubing but not fluid movement.

油管锚实质上是没有密封作用的封隔器，目的是防止油管而不是流体移动。

surface choke *np.* 地面油嘴

例：Generally, the most important completion components are the tubing string and the surface choke, because these components usually have the greatest effect on the flowing performance of a well.

一般来说，最重要的完井设备是油管柱和地面油嘴，这些设备通常对油井流动性影响最大。

casing–tubing annulus *np.* 套管—油管环空/油套环形空间

例：The packer seals the casing-tubing annulus with a rubber packing element, thus preventing flow and pressure communication between tubing and annulus.

封隔器用橡胶材料密封油套环形空间，防止油管和环空间流体和压力串通。

multistring packer *np.* 多管封隔器

例：The multistring packer seals the casing-tubing annulus where more than one tubing string is involved.

多管封隔器用于多个油管情况下油套环空的封隔。

mechanically set packers *np.* 机械坐封封隔器

例：Mechanically set packers rely on tubing or drillpipe movement to force grooved "slips" to grip the casing and to expand the sealing element during the setting procedure.

在坐封过程中，机械坐封封隔器依靠油管或者钻杆运动使得带槽卡瓦紧扣套管，并膨胀密封元件。

casing hanger *np.* 套管悬挂器

例：All wellheads have at least one casing head and <u>casing hanger</u>, usually, a tubing head and tubing hanger, and a Christmas tree.

井口至少都有一个套管头和套管悬挂器，通常井口都有油管头、油管悬挂器以及采油树。

casing spool *np.* 套管防喷法兰短节

例：Normally, the casing head and <u>casing spools</u> have at least one additional connection designed to allow fluid access and pressure monitoring of the concentric annular spaces during production.

正常情况下，套管头和套管四通至少会有一个附加连接，以允许流体进入以及生产期间对环空进行压力监测。

throttling pressure valve *np.* 扼流压力阀

例：A <u>throttling pressure valve</u> is sensitive to tubing pressure in the open position, and once opened by casing pressure buildup, requires a reduction in tubing or casing pressure to close.

扼流压力阀对安装位置的油管压力很敏感，一旦由于套管压力升高而被打开，需要降低油管或套管压力才能关闭。

sliding/circulating sleeve *np.* 滑套

例：The <u>sliding sleeve</u> component is a wireline-operated sleeve, which will open or close ports in the tubing to allow fluid in or out.

滑套组件是一个由钢丝绳控制的套筒，可以打开或关闭管道中的端口，以允许流体进出。

blast joint *np.* 耐磨接头

例：A <u>blast joint</u> is a section of heavy-duty tubing located opposite production perforations in a multistring completion.

在多管柱完井中，耐磨接头是重型油管的一部分，位于生产射孔对面。

subsurface safety valve *np.* 井下安全阀

例：A subsurface safety valve is installed in the tubing string near the surface.

井下安全阀安装在靠近地面的油管柱中。

bottom/master valve *np.* 底阀

例：In the Christmas tree, the bottom valve, often called the master valve, is the primary means for completely shutting in the well.

在采油树中，底阀通常称为主阀，是实现完全关闭油井的主要手段。

standing valve *np.* 定阀

例：The barrel, plunger and standing valve of the pump were designed, improving the barrel's force conditions and reinforcing the screw thread's sealing performance and tensile strength.

该泵的泵筒、柱塞、定阀均做了一定设计，改善了泵筒受力状况，提高了螺纹的密封性能和抗拉强度。

primary recovery *np.* 一次采油

例：The oil obtained as a result of the natural production mechanisms is referred to as "primary recovery" where oil recoveries could have been 20 per cent or less.

通过天然采油机理开采石油称为"一次采油"，其采收率最多达到百分之二十。

primary phase of production *np.* 一次采油阶段

例：The field is said to be in the primary phase of production as long as there is sufficient pressure left in the reservoir to bring the oil to the bottom of the producing wells.

只要油田的储油层有足够的压力，能够把石油带到生产井井底，这样的油田就处于一次采油阶段。

natural flowing/natural lift *np.* 自喷采油

例：In natural lift, as long as the pressure under which the oil exists in the reservoir is greater than the pressure at the bottom of the well, the oil will continue to flow into the well bore.

在自喷采油中只要油藏压力大于井底压力，石油就会源源不断地流入井筒。

reservoir pressure *np.* 油藏压力

例：The rate at which the oil will move through the rock towards the well will depend on the reservoir pressure, the pressure at the bottom of the well, the permeability of the rock and the viscosity of the oil.
石油通过岩石流向井里的速度取决于油藏压力、井底压力、岩石的渗透率和石油的黏度。

displacing action *np.* 驱替作用

例：The displacing action of the water will result in the recovery of a relatively high proportion of the oil originally present in the oil zone.
水的驱替作用将油层中的油更多地开采出来。

drive mechanism *np.* 驱动方式

例：The three principal primary recovery drive mechanisms are water drive, gas drives, and gravity drainage.
一次采油阶段主要有三种油藏驱动方式，分别是水驱、气驱和重力泄油。

water drive *np.* 水驱

例：Water drive uses the pressure exerted by water below the oil and gas in the formation to force hydrocarbons out of the reservoir.
水驱利用油气层下方水所施的压力，将油气驱出储层。

edgewater drive *np.* 边水驱动

例：Edgewater drive—the aquifer exclusively feeds one side or flank of the reservoir.
边水驱动是指含水层在油藏的单侧或两侧为油层提供驱动力。

bottomwater drive *np.* 底水驱动

例：Bottomwater drive—he aquifer underlays the reservoir and feeds it from beneath.
底水驱动是指含水层位于油藏下方，为油层提供驱动力。

compaction drive *np.* 压实驱动

例：Compaction drives characteristically exhibit elevated rock compressibilities, often 10 to 50 times greater than normal.
压实驱动的特点表现为岩石压缩性提高，通常是正常岩石的10到50倍。

water influx *np.* 水侵

例：The degree to which water influx improves oil recovery depends on the size of the adjoining aquifer, the degree of communication between the aquifer and petroleum reservoir, and ultimately the amount of water that encroaches into the reservoir.

水侵能在多大程度上提高采收率，取决于相邻含水层的大小、含水层与油藏的连通程度以及最终侵入油藏的水量。

salt water *np.* 盐水

例：The aquifer is under pressure and the salt water contained in it is one source of potential energy.

含水层承受着一定的压力，其中的盐水是一种潜在能源。

fingering *n.* 指进

例：The phenomenon is known as "fingering" when the encroaching aquifer water travels through the more permeable sections of the rock, by-passing the less permeable ones.

当含水层侵入的水绕开渗透率较低的地段，而从岩层中渗透率较高的地段通过时，这种现象就是"指进"。

water coning *np.* 水锥

例：If a well is completed too near the oil /water level, its economic life may be short as it may soon become "watered out". This phenomenon is known as "water coning".

如果完井层位过于接近油水界面，很快就会出现"水淹"，经济寿命可能十分短暂。这种现象称为"水锥"。

free gas *np.* 游离气

例：Some free gas or water may flow into the well with the oil and these additional fluids tend to reduce the amount of oil produced by the well.

一部分游离气或游离水可能会随油一起流入井里，这些流体流出可能会减少该油井产油量。

saturation pressure *np.* 饱和压力

例：As the production of oil progresses, the pressure of the whole reservoir will fall below <u>saturation pressure</u> and at this point some of the gas which then comes out of solution will find its way upwards inside the reservoir to form a secondary gas cap.

随着石油开采，油藏压力会降到饱和压力之下，此时一些溶解气会出来窜到油藏上部形成次生气顶。

gas drive *np.* 气驱

例：The two types of <u>gas drives</u> are dissolved-gas drives and gas-cap drives. These drives use the pressure of gas in the reservoir to force oil out of the reservoir and into the well.

气驱分为溶解气驱和气顶驱动，其工作原理是利用储层中的气体压力驱替石油从储层进入油井中。

dissolved gas *np.* 溶解气

例：As long as the difference in pressure can be maintained, the oil with its associated <u>dissolved gas</u> will continue to flow into the well bore.

只要这种压力差能够维持，石油及其伴随的溶解气就会源源不断地流入井筒。

dissolved gas drive/solution gas drive *np.* 溶解气驱

例：<u>Solution gas drive</u> is by no means as effective as water drive or gas cap drive and a smaller proportion of the total oil in the reservoir is recovered by this method.

溶解气驱远不如水驱或气顶驱那么有效，用这种方法只能开采出油藏中很少一部分石油。

secondary gas cap *np.* 次生气顶

例：<u>Secondary gas cap</u> will assist production in the same manner as the gas cap drive mechanism.

和气顶驱机理一样，次生气顶也有助于采油。

gas cap *np.* 气顶

例：As oil is withdrawn from the reservoir and the pressure drops, the <u>gas cap</u> expands to fill the pore space formerly occupied by the oil, and thus displaces oil downwards towards the producing well.

当石油从油藏中排出而压力下降时，气顶膨胀并充斥之前石油占据的孔隙空间，并且向下驱替石油至生产井。

gas coning *np.* 气锥

例：A high production rate in a well located near the gas/oil level may cause "gas coning".

靠近油气界面的井，如果采油速度很快，可能导致"气锥"。

gravity drainage *np.* 重力泄油

例：When the energy of a solution gas drive or gas cap drive production mechanism has largely been expended, the dominant cause of production is the drainage of oil into the well bore under its own gravity.

当溶解气驱或气顶驱的能量基本上消耗殆尽时，石油就要依靠自身重力作用泄入井内。

steeply dipping reservoir *np.* 陡斜油藏

例：Steeply dipping reservoir are classic examples of where gravity drainage provides an effective recovery mechanism.

重力泄油适用于陡斜油藏，是一种有效采油方式。

production rate *np.* 产量

例：The production rate has to be carefully controlled to prevent undue waste of gas.

必须注意控制产量，防止天然气白白浪费。

secondary recovery *np.* 二次采油

例：The oil which is forced out of the reservoir rock as a result of the application of an artificial technique is known as "secondary recovery".

应用人工技术从储油岩里驱替石油称为"二次采油"。

artificial lift *np.* 人工举升

例：The insufficiency of reservoir natural pressure to force oil out of the pores of the rock led to the development of artificial lift to force more oil out of the pore spaces of the reservoir rock.

当油藏压力不足以将岩石孔隙里的油驱替出来时，人工举升法应运而生，目的是从储油岩孔隙中驱替出更多的石油。

gas injection *np.* 注气

例：Artificial lift uses pumps and gas injection.

人工举升通常使用泵抽和注气。

continuous gas lift *np.* 连续气举

例：Continuous gas lift relies on the constant injection of gas-lift gas into the production stream through a downhole valve.

连续气举通过井下阀门将气体持续注入采出液。

intermittent gas lift *np.* 间歇气举

例：Intermittent gas lift allows for the buildup of a liquid column of produced fluids at the bottom of the wellbore. At the appropriate time, a finite volume of gas is injected below the liquid and propels it as a slug to the surface.

间歇气举可以在井筒底部形成产液柱，并在适当时候向液柱下方注入一定量气体，形成段塞，从而排出井中液体。

plunger lift *np.* 柱塞举升

例：Plunger lift is typically an intermediate artificial lift method for wells that ultimately must be pumped but have a low productivity index (PI) and a high enough gas-oil ratio to operate the plunger.

柱塞举升是一种典型的人工举升方法，适用于采油指数低但又必须泵油的油井，其前提是气油比够高，足以使柱塞工作。

chamber lift *np.* 箱式气举

例：Chamber lift, a gas-lift installation, which allows for production from low PI wells without the backpressure from injected gas.

箱式气举也是一种气举采油方法，不会因注气产生反压，适用于采油指数低的油井。

artificial lift equipment *np.* 人工举升设备

例：Artificial lifting equipment refers to the devices needed to increase the flow rate of a production well.

人工举升设备是指提高生产井产量所需的设备。

gas-lift valve *np.* 气举阀

例：A type of gas-lift valve that allows a gas-lift port size to be adjusted remotely from surface to positions from fully open to closed. These valves offer the possibility of changing gas-injection points without well intervention.

气举阀可以从地面远程调节气举端口的大小，从全开到关闭。这些阀门使无需油井干预即可改变注气点成为可能。

sucker rod pumping (SRP) *np.* 抽油杆泵

例：The most common is sucker rod pumping, where the pumping motion is transmitted from the surface to the pump by means of a string of narrow jointed rods placed within the tubing.

抽油杆泵最常用，其动力是通过放置在油管内的一系列窄连接杆从地面传下来。

beam pumping *np.* 梁式抽油

例：Sucker rod or beam pumping is the most common method (85%), with gas lift second (10%), and then electrical submersible and hydraulic pumping about equal (2%) in usage.

有杆泵或梁式抽油是最常用的采油方式，占比约85%；其次是气举，占比10%，最后是电潜泵和液压泵，各占2%左右。

beam pumping unit *np.* 游梁抽油机

例：The beam pumping unit sits on the surface and creates an up-and-down motion to a string of rods called sucker rods.

游梁抽油机位于地面，带动一组抽油杆进行上下活塞运动。

hydraulic pumping unit (HPU) *np.* 液压抽油机

例：HPU (hydraulic pumping unit) is one type of sucker rod pump (SRP). Hydraulic pump is used as an alternative to the artificial lift system.

液压抽油机是一种有杆泵（SRP），属于人工举升法。

extraction pump *np.* 抽油泵

例：The treatment equipment comprises a waste oil tank, an extraction pump, an oil filter, an impurity separator, a pressuring pump and a ceramic film filter.

处理设备包括废油槽、抽油泵、滤油器、杂质分离器、加压泵和陶瓷膜过滤器。

progressive cavity pump *np.* 螺杆抽油泵

例：Progressive cavity pumps as oil transfer pumps have been used for oil transfer over the last few years because of a number of benefits they offer.

螺杆抽油泵可提供许多便利，过去几年间一直用于抽油。

reciprocating plunger pump *np.* 往复式柱塞泵

例：Reciprocating plunger pumps are among the most common types of pumps used in the oil and gas industry.

在石油和天然气行业中，往复式柱塞泵是最常见的一种泵。

tubing pump *np.* 管式泵

例：In beam pump system, we usually use tubing anchor to fix pump, and the bleeder installed on the tubing pump to improve the working environment.

在梁氏泵系统中，经常采用油管锚固定泵，通过在管式泵的上方安装泄油器来改善油井施工作业时的工作环境。

beam pump *np.* 梁式泵

例：As the widely used equipment in mechanical oil withdrawal field, beam-pump unit's total efficiency is lower than 25%.

作为机械采油领域广泛使用的设备，梁式泵的总效率低于25%。

rodless pump *np.* 无杆泵

例：The majority of rodless subsurface pumps fall into two categories: hydraulic and electrical submersible centrifugal.

无杆泵主要分为液压式和电潜式。

electrical submersible pump (ESP) *np.* 电潜泵

例：The electrical submersible pump, typically called an ESP, is an efficient and reliable artificial-lift method for lifting moderate to high volumes of fluids from wellbores.

电潜泵（ESP）是一种高效、可靠的人工举升方法，可从井筒中举升较大量的石油。

progressing cavity pump (PCP) *np.* 螺杆泵

例：Progressing cavity pump (PCP) systems derive their name from the unique, positive displacement pump that evolved from the helical gear pump concept first developed by Rene Moineau in the late 1920s.

螺杆泵（PCP）系统得名于一种独特的容积式泵，这种容积式泵是由20世纪20年代Rene Moineau首次提出的螺旋齿轮泵概念演变而来的。

jet pump *np.* 射流泵

例：These include: jet pumping, a hydraulic pump, which uses a nozzle to transfer power fluid momentum directly to the produced fluid.

其中包括射流泵，一种液压泵，使用喷嘴将流体动量直接传递给采出液。

static fluid level *np.* 静液面

例：Static fluid level is the level to which fluid rises in a well when the well is shut in.

静液面是关井后环空液面缓升到一定位置稳定下来的液面。

actual pump displacement *np.* 实际泵排量

例：Volumetric efficiency reflects the relationship between actual pump displacement and the pump displacement under ideal conditions.

实际泵排量与理想泵排量之间的关系通过容积效率呈现。

water flooding *np.* 注水开发

例：Water flooding is most effective in reservoirs in which the permeability is fairly uniform and in such cases oil recoveries of some 40 percent or more of the oil in place are now common.

在渗透率比较均匀的油藏里注水采油效果最好，目前采收率一般可达约百分之四十以上。

injection well *np.* 注入井

例：Water is injected under pressure into reservoir rocks via injection wells, thus driving the oil through the rock towards the producing wells.

高压水经过注水井注入储油岩，驱动石油穿过岩石流向生产井。

water displacement *np.* 水驱替

例：Water displacement is a particular case of fluid displacement, which is simply the principle that any object placed in a fluid causes that fluid to no longer occupy that volume of space.

水驱替是流体驱替的一种特殊情况，其原理是向油藏注入水或气，使其挤压流体的体积空间，从而驱动油气喷出。

displacing fluid *np.* 驱替液

例：The displacing fluids must be appreciably cheaper than the oil which they force out of the pores.

驱替液成本必须大幅低于其从孔隙中驱替的石油成本。

dry gas *np.* 干气

例：If the dry gas has a ready sale, the reservoir pressure can be maintained by water injection.

如果干气畅销，可以改用注水的办法维持气藏压力。

cycling the produced gas *np.* 干气回注

例：The reservoir pressure can be maintained by re-injecting or "cycling" the produced gas back into the reservoir after the condensate liquid has been extracted at the surface.

通过"干气回注"可以保持气藏压力，即把地面上脱去凝析液的气体再度注入气藏。

nitrogen injection *np.* 注氮气

例：Nitrogen injection is used to enhance oil recovery in many oilfields, particularly in overthrust fault.

目前许多油田，特别是逆掩断层油藏，用注氮气法提高原油采收率。

carbon dioxide injection *np.* 注二氧化碳

例：The recovery of coalbed methane can be efficiently enhanced by carbon dioxide injection.

注二氧化碳可以有效提高煤层气的采收率。

air injection *np.* 注空气

例：Now air injection into light oil for EOR has received more attention.

通过给轻质原油注空气来提高采收率，该方法目前受到更广泛关注。

flue gas injection *np.* 注烟道气

例：With the increase of reservoir temperature, the displacement efficiency of flue gas injection is higher than that of air injection.

随着储层温度的升高，注烟道气的驱油效率大于注空气。

steam injection *np.* 注蒸汽

例：Hot water injection is less effective than steam injection since steam has a higher heat content than hot water.

注热水的效果不如注蒸汽，因为蒸汽的热含量比热水高。

steam soak/steam huff and puff *np.* 蒸汽吞吐

例：Steam huff and puff injection is one of the thermal EOR methods in which steam is injected in a cyclical manner alternating with oil production.

蒸汽吞吐是热力采油的方法之一，注蒸汽与采油周期性交替进行。

steam assisted gravity drainage (SAGD) *np.* 蒸汽辅助重力泄油

例：While the retrofit system is applicable to most production scenarios, it has found the most applications to date in steam assisted gravity drainage (SAGD) wells.

改造后系统虽然适用于大多数生产场合，但迄今为止，在蒸汽辅助重力泄油（SAGD）井中的应用最多。

cyclic steam stimulation (CSS) *np.* 循环注气

例：Cyclic steam stimulation is a method of thermal recovery in which a well is injected with steam and then subsequently put back on production.

循环注气是一种热采方法，即注蒸汽到油井中并继续开采。

improved oil recovery(IOR) *np.* 改进原油采收率

例：The IOR methods are also expected to reduce this production gap by bringing advanced technological solutions to boost oil recovery from mature fields, geologically complex low permeable and carbonate reservoirs in the future.

IOR 方法还将通过引入先进的技术解决方案，提高成熟油田、地质条件复杂的低渗透和碳酸盐岩储层的原油采收率，从而缩小产量差距。

recovery factor *np.* 采收率

例：Recovery factor is the percentage of OIP or GIP that the reservoir will produce.

油藏的采收率是指产油/气量与地质储量之比，一般用百分数表示。

enhanced oil recovery (EOR) *np.* 提高采收率

例：Enhanced oil recovery is also called improved oil recovery or tertiary recovery, the third stage of hydrocarbon production during which sophisticated techniques that alter the original properties of the oil are used.

提高采收率也称三次采油，即在油气开采第三阶段使用复杂技术改变原油特性。

ultimate recovery *np.* 最终采收率

例：The fracture azimuth will change the flood response time, and then change the ultimate recovery.

裂缝方位角会改变水驱响应时间，进而改变最终采收率。

heavy oil recovery *np.* 稠油开采

例：Thermal recovery is the most effective method for heavy oil recovery.

热力采油是稠油开采最有效的方法。

thermal production *np.* 热采

例：Thermal production includes injecting thermal fluid into reservoir, such as cyclic steam stimulation (CSS), steam flooding (SF), and steam assisted gravity drainage (SAGD) or generating heat in the reservoir.

热采指的是向油藏中注入热流体，如蒸汽吞吐（CSS）、蒸汽驱（SF）、蒸汽辅助重力泄油（SAGD）或在油藏中产热。

thermal recovery *np.* 热力驱

例：One of the most efficient approaches for enhanced oil recovery (EOR) in heavy oil reservoirs is thermal recovery.

稠油油藏提高采收率（EOR）最有效的方法之一是热力驱。

in situ combustion/underground combustion *np.* 火烧油层

例：The heating effect achieved by burning underground some of the oil in the reservoir is referred to as "in situ combustion", in which air is injected to support combustion and to act as the displacing medium.

在地下点燃油层中一部分原油来加热油藏就是所谓的"火烧油层"，作业时注入空气，空气可以助燃，同时充当驱替介质。

sand face ignition *np.* 油层点火

例：In situ combustion is an important method for thermal recovery of viscous oil reservoir, the key of success is sand face ignition.

火烧油层是热力开采稠油油藏的重要方法，其成功的关键是油层点火。

steam drive *np.* 蒸汽驱

例：Steam drive in heavy oilfield development is on its beginning in our country.

稠油油田蒸汽驱采油在我国处于起步阶段。

hot water flooding *np.* 热水驱

例：Hot water flooding is a thermal recovery technique in which hot water is injected into a reservoir to increase heavy to medium crude oil production.

热水驱是一种向油藏注入热水以提高稠油和中质原油产量的热采技术。

cold production *np.* 冷采

例：Cold production methods include chemical flooding (polymer flooding, surfactant flooding, gas flooding etc.), natural depletion and water flooding.

冷采方法包括化学驱（聚合物驱、表面活性剂驱、气驱等）、自然衰竭和水驱。

chemical flooding *np.* 化学驱

例：Three major types of EOR operations are chemical flooding, miscible gas drive and thermal recovery.

提高采收率的三种主要方式是化学驱、混相气驱和热力驱。

miscible gas drive *np.* 混相气驱

例：The difficulty of recovering oil from reservoirs containing heavy viscous crudes led to the development of miscible gas drive techniques.

从高黏度重质油油藏中开采原油很困难，因此，混相气驱技术发展起来。

partially miscible displacement *np.* 部分混相驱替

例：An enhanced oil recovery method in which carbon dioxide (CO_2) is injected into a reservoir to increase production by reducing oil viscosity and providing miscible or partially miscible displacement of the oil.

提高采收率的方法之一是通过向储层注入二氧化碳，以此降低石油黏度，进行混相或部分混相石油驱替，以增加产量。

alkaline flooding *np.* 碱水驱

例：Variation of interfacial tension (IFT) in alkaline flooding process is studied through a series of experiments.

通过一系列试验，研究了碱驱过程中界面张力的变化规律。

carbonated water flooding (CWF) *np.* 碳酸水驱

例：Carbonated water flooding (CWF) appears to be an important means in enhanced oil recovery (EOR) in carbonate reservoirs.

碳酸水驱是碳酸盐岩油藏提高采收率的一种重要手段。

polymer flooding *np.* 聚合物驱

例：Polymer flooding technology enhances oil recovery by increasing the sweep efficiency of water.

聚合物驱油技术是通过提高水的波及系数来提高原油采收率。

enriched gas drive *np.* 富气驱

例：Enriched gas drive is a miscible oil flooding technology which uses rich gas as miscible injection agent.

富气驱是一种以富气为混相注入剂的混相驱油技术。

surfactant flooding *np.* 表面活性剂驱

例：Surfactant flooding is an Enhanced Oil Recovery technique in which the mobility of residual oil in the reservoir is increased by reducing the Interfacial Tension (IFT) between the injected fluid and the reservoir oil.

表面活性剂驱是一种提高采收率的技术，通过降低注入流体与油藏原油之间的界面张力（IFT）来提高油藏中剩余油的流动性。

surface tension *np.* 表面张力

例：The chemicals are designed to reduce the surface tension of the remaining oil and wash it to a producing well.

这些化学物质旨在降低剩余油的表面张力，并将剩余油驱至生产井。

foam drainage *np.* 泡沫排水

例：Foam drainage technology has been widely used to remove the liquid out of pipe in loading gas wells and keep the well normally producing.

泡沫排水技术已广泛应用于储气井管道抽汲，以保证气井的正常生产。

microbial oil recovery (MOR) *np.* 微生物采油

例：At present, the studies on microbial oil recovery are mainly in single-strain microbial oil recovery.

目前微生物采油技术的研究主要集中在单菌种微生物提高采收率方面。

gas recovery *np.* 天然气开采

例：Gas recovery by water injection is, unlike oil, not improved sufficiently by water displacement to justify the expense of such an operation.

和注水驱油情况不同，注水开采天然气的做法尚不完善，结果往往得不偿失。

gas deliver-ability test *np.* 天然气产能测试

例：Modified isochronal test is a type of gas deliver-ability test conducted in gas wells to generate a stabilized deliver-ability curve (IPR).

修正等时测井是在气井中进行的一种天然气产能测试，以生成稳定的产能曲线。

gas-bearing rock *np.* 含气岩层

例：Once the well has been drilled, the final casing cemented in place across the gas-bearing rock has to be perforated in order to establish communication between the rock and the well.

钻井结束后，必须在含气岩层上进行最后一层固井套管的射孔，以便在岩石和井之间建立连通。

cracking/pyrolysis gas *np.* 裂解气

例：The cracking gas of reservoir oil is an important source of natural gas.

油藏原油的裂解气是形成天然气气藏的重要气源。

retrograde condensation *np.* 反转凝析

例：If the reservoir is being produced for the sake of its condensate only it is important to prevent retrograde condensation taking place in the reservoir.

如果开采气藏仅仅为了获得凝析油，那么防止反转凝析发生很重要。

condensate *n.* 凝析油

例：This liquid does not come from a separate oil zone, but exists as a gas in the reservoir and is made up of that part of the reservoir gas which condenses as its pressure is reduced by production. The liquid fraction is known as "condensate".

这种液态石油并非产自独立的油层，而是呈气态存在于气藏中的。气藏中的这部分天然气，在生产过程中因压力下降而凝结，成为液态即"凝析油"。

venting *n.* 排气

例：Venting and/or flaring of the gas at this stage are the main reasons why shale and tight gas can give rise to higher greenhouse-gas emissions than conventional production.

该阶段排放或燃烧气体是导致页岩气和致密气生产中温室气体排放量比常规油气生产高的主要原因。

flaring *n.* 点火处理/燃烧

例：In these cases, there can be venting of gas to the atmosphere (mostly methane, with a small fraction of volatile organic compounds) or flaring (burning) of hydrocarbon or hydrocarbon/water mixtures.

鉴于此类情况，可以将气体排放到大气中（其主要成分是甲烷，也含有小部分挥发性有机化合物），或对烃或烃／水混合物进行点火处理。

oil displacement efficiency *np.* 驱油效率

例：Experimental results show that the oil displacement efficiency model is determined by many factors together.
实验结果表明，驱油效率模型是由多个因素共同决定的。

formation permeability and fluidity *np.* 储层渗流

例：The tectonic fractures formed latterly improve the formation permeability and fluidity, and determine reservoir production capability.
后期形成的构造裂缝改善了储层的渗流特性，决定储层产液能力。

bottom hole back pressure *np.* 井底回压

例：Techniques of gas recovery by water drainage with liquid foamer can reduce the bottom-hole pressure, improve the water-carrying ability of gas flow, and increase the gas production.
液体起泡剂排水采气工艺技术可降低井底回压，提高气流携水能力，从而达到增加天然气产量的目的，从原理性能上优于常规投棒排水技术。

very high water cut *np.* 特高含水

例：As an old offshore oilfield with a production history over 33 years, CR has reached a stage of very high water cut, with its recovery degree more than 40%.
CR 油田是经历了 33 年生产历史的海上老油田，目前已进入特高含水期，采出程度超过 40%。

stripper well *np.* 低产井

例：This apparatus is mainly used in oil extracting operation of margin well, stripper well and abandon well of oil field.
该设备主要用于对油田边缘井、低产井及废弃井的采油作业。

plug removal *np.* 解堵

例：Plug removal of oil well using high pressure water jet is a new developing technology in recent years.

利用高压水射流技术进行油井解堵，是近年来发展的一种新技术。

wax removal *np.* 清蜡

例：Heat wash of the oil well is the most commonly used method of wax removal.
油井热洗是油田最常用的一种清蜡方法。

Section III

Unconventional Oil and Gas Extraction
非常规油气生产

unconventional field *np.* 非常规油气田

例：Whereas onshore conventional fields might require less than one well per ten square kilometers, unconventional fields might need more than one well per square kilometer (km^2), significantly intensifying the impact of drilling and completion activities on the environment and local residents.

陆上常规油田每十平方公里最多需要一口井，而非常规油气田每一平方公里可能就至少需要一口井，这就使钻井和完井工程对环境和当地居民的影响大幅增加。

unconventional resource *np.* 非常规资源

例：Unconventional resources are less concentrated than conventional deposits and do not give themselves up easily.

非常规资源比常规资源的沉积更加分散，而且不易开采。

unconventional gas *np.* 非常规天然气

例：Unconventional gas refers to a part of the gas resource base that has traditionally been considered difficult or costly to produce.

非常规天然气指一直被认为开发难度大、成本高的天然气资源。

shale gas *np.* 页岩气

例：Shale gas is natural gas contained within a commonly occurring rock classified as shale.

页岩气是蕴藏在页岩（一种常见岩石）中的天然气。

coalbed methane *np.* 煤层气

例：<u>Coalbed methane</u>, also known as coal seam gas, is natural gas contained in coalbeds.

煤层气，也称为煤层甲烷，是蕴藏在煤层中的天然气。

tight gas *np.* 致密气

例：<u>Tight gas</u> is a general term for natural gas found in low permeability formations.

致密气是对低渗透率地层中天然气的通称。

shale formation *np.* 页岩层

例：<u>Shale formations</u> are often rich in organic matter and, unlike most hydrocarbon reservoirs, are typically the original source of the gas.

页岩层富含有机质，与大部分油藏不同的是，页岩地层是典型的气源岩。

low permeability gas reservoir *np.* 低渗透气藏

例：These <u>low permeability gas reservoirs</u> cannot produce economically without the use of technologies to stimulate flow of the gas towards the well, such as hydraulic fracturing.

只有使用水力压裂等增产措施使气体加速流向井底，这些低渗透气藏生产才经济可行。

gas shale *np.* 气页岩

例：Shale formations are therefore viewed as both source rocks and reservoirs for natural gas and are often called <u>gas shales</u>.

页岩地层既是天然气的气源岩又是储集岩，常称为气页岩。

lamination *n.* 层状结构

例：The combination of compaction and <u>lamination</u> result in the shale formation with extremely limited permeability.

压实作用和层状结构导致页岩层渗透率极低。

matrix permeability *np.* 基质渗透率

例：The low <u>matrix permeability</u> of shale means that gas trapped cannot move easily within the rock.

页岩基质渗透率较低，因此岩石内聚集的气体无法轻易流动。

associated hydrocarbon *np.* 伴生的碳氢化合物

例：Natural gas is considered "dry gas" when it is almost pure methane, having had most of the other commonly associated hydrocarbons removed.

将大部分伴生碳氢化合物去除后，天然气几乎为纯甲烷，称为"干气"。

wet gas *np.* 湿气

例：Although some shale formations do produce wet gas, shale gas is typically a dry gas primarily composed of 90 percent or more methane.

尽管一些页岩地层也会产生湿气，但页岩气一般为干气，其甲烷含量在90%以上。

nonassociated gas *np.* 非伴生气

例：Nonassociated gas comes from reservoirs that are not connected with any known source of liquid petroleum.

非伴生气来自与任何已知液态石油来源都无联系的储层。

associated gas *np.* 伴生气

例：Associated gas is produced during crude oil production and is the gas that is associated with crude oil.

伴生气是与原油伴生的，在原油开采过程中一同采出。

economically viable basin *np.* 经济可采盆地

例：The specific characteristics of gas shales have been found to vary both among and within economically viable basins.

在经济可采盆地内及盆地之间，气页岩都有其各自特性。

horizontal pay streak *np.* 水平产层

例：These technologies enable wells to reach and accurately follow long, narrow, horizontal pay streaks.

这些技术可以使气井到达产层并沿狭长的水平产层精确地延伸。

reservoir stimulation technique *np.* 油藏增产措施

例：Promoting gas flow from low-permeability shales has been further facilitated by advances in reservoir stimulation techniques, specifically hydraulic fracturing.

油藏增产措施（特别是水力压裂）的发展，进一步促进了低渗透页岩中气体的流动。

formation fracturing *np.* 地层压裂

例：Lithology, pore development is a key factor to affect the formation fracturing operation.

岩性和孔隙发育是影响地层压裂作业的关键因素。

oil-based fracturing fluid *np.* 油基压裂液

例：Oil-based fracturing fluid and foam-based fracturing fluids using bubbles of nitrogen or carbon dioxide can be used to minimize formation damage.

油基压裂液和利用二氧化碳气或氮气泡的泡沫压裂液，能最大程度减少对地层的伤害。

target rock *np.* 目标岩层

例：Hydraulic fracturing involves pumping a fluid known as fracturing fluid at high pressure into the well and then into the surrounding target rock.

水力压裂技术的原理是将压裂液高压泵入井内，进入周围的目标岩石。

high energy gas fracturing *np.* 高能气体压裂

例：The fracturing technology with high energy gas is a new stimulation developed in1980s.

高能气体压裂技术是20世纪80年代发展起来的一种油气井增产新技术。

flow-back period *np.* 返排期

例：During this flow-back period, typically over days (for a single-stage fracturing) to weeks (for a multistage fracturing), the amount of flow back of fracturing fluid decreases, while the hydrocarbon content of the produced stream increases, until the flow from the well is primarily hydrocarbons.

返排期通常会持续数天（一段压裂）到数周（多段压裂）。在此期间，排出的压裂液越来越少，而采出流体中的烃含量逐渐增加，直到井流基本上都是烃类物质。

acid fracturing *np.* 酸化压裂

例：<u>Acid fracturing</u> is an important technology for oil field production and augmented injection.

酸化压裂是油田生产和增注的一项重要技术。

acid treatment *np.* 酸化处理

例：<u>Acid treatment</u> involves the injection of small amounts of strong acids into the reservoir to dissolve some of the rock minerals and enhance the permeability of the rock near the wellbore.

酸化处理技术是向储层注入少量的强酸来溶解部分岩石矿物，从而提高井筒附近岩石的渗透率。

processing facilities *np.* 生产设备

例：Once wells are connected to <u>processing facilities</u>, the main production phase can begin.

一旦油气井与生产设备连通，即进入主要生产阶段。

coal seam *np.* 煤层

例：As with shale and tight gas deposits, there are major variations in the concentration of gas from one area to another within the <u>coal seams</u>.

与页岩气和致密气藏相似的是，煤层中各处的天然气含量不同。

honing technique *np.* 珩磨技术

例：Considerable progress has been made over the last 25 years in <u>honing techniques</u> to extract coalbed methane on a commercial basis.

过去25年来，珩磨技术用于煤层气商业化开采取得了重大进步。

dewatering *n.* 气井排水

例：Once a well is drilled, the water in the coalbed is extracted, either under natural pressure or by using mechanical pumping equipment—a process known as <u>dewatering</u>.

井钻好后，就要利用自然压力或机械泵将煤层中的水采出，这一过程叫作气井排水。

cleat/joint *n.* 节理

例：As subsurface pressure drops with dewatering, the flow of natural gas previously held in place by water pressure increases initially as it is released from the natural fractures or <u>cleats</u> within the coalbed.

地下压力随着水排出而下降，原来因水压而滞留的天然气从煤层的天然裂隙或节理中释放出来，加速流动。

lined containment pit/tank *np.* 防渗漏的储存池或储存罐

例：The flow-back fluids recovered from the well are pumped to <u>lined containment pits</u> or tanks for treatment or offsite disposal.

井中返排流体要泵入防渗漏的储存池或储存罐中，供后续处理或排放。

impermeable layer *np.* 不透水层

例：Leakage through the rock from the producing zone is unlikely in the case of shale gas or tight gas because the producing zone is one to several thousand metres below any relevant aquifers and this thickness of rock usually includes one or several very <u>impermeable layers</u>.

产层油气穿过周围岩层泄漏这种现象在页岩气和致密气开采中不可能出现，因为这两种气藏的产层位于含水层以下一千至几千米深的地方，产层和含水层之间通常由一个或多个渗透性极差的岩层隔开。

spills at the surface *np.* 地表溢流

例：The first hazard—the risk of <u>spills at the surface</u>—can be mitigated through rigorous containment of all fluid and solid streams.

第一大风险是地表溢流风险，严格密闭液相和固相流体可降低这类风险。

polymer *n.* 聚合物

例：Gelling <u>polymers</u> are used in the fracturing fluid at a concentration of about 1%.

胶凝聚合物在压裂液中使用比例约为1%。

formulation *n.* 配方

例：The number, type, and concentration of chemical additives used in fracture treatments vary depending on specific well conditions and <u>formulations</u> are generally proprietary to the fracturing fluid producer.

用于压裂处理的添加剂数量、类型和浓度要根据具体井况和地层条件而定，其配方通常为压裂液生产商所专有。

productive life *np.* 开采寿命

例：For an unconventional development, the productive life of a well is expected to be similar, but shale gas wells typically exhibit a burst of initial production and then a steep decline, followed by a long period of relatively low production.

至于非常规油气开发，油气井的开采寿命预计与常规井差不多，但页岩气井通常初始产量高，随后产量急剧下降，然后在相当一段时间内产量很低。

inflow control devices (ICDs) *np.* 流量控制技术

例：Using inflow control devices (ICDs) or autonomous inflow control devices (AICDs) in conjunction with sand screens can maximize recovery by maintaining sand control integrity and effective drainage.

将流量控制技术（ICDs）或自动流入控制装置（AICDs）与防砂筛管结合使用，可以完全防砂，还能高效排水，从而最大程度提升采收率。

shale oil *np.* 页岩油

例：Unconventional oil consists of more liquid sources including oil sands, extra heavy oil and shale oil, which need advanced technology to be extracted.

非常规石油种类更多，包括油砂、超重石油和页岩油，这些油需要先进的技术才能采出。

extracting oil *np.* 采油

例：Tar sands are mined and processed to generate oil similar to oil pumped from conventional oil wells, but extracting oil from tar sands is more complex than conventional oil recovery.

焦油砂开采出的油与常规油井采出的油相似，但前者的开采过程比后者更加复杂。

tar sand extraction *np.* 油砂开采

例：Tar sand extraction is more complex than conventional oil recovery.

油砂开采比常规石油开采更加复杂。

Part III Oil and Gas Exploitation 油气开发

tar sand *np.* 焦油砂

例：Tar sands (also referred to as oil sands) are a combination of clay, sand, water, and bitumen, a heavy black viscous oil.

焦油砂，也称为油砂，是一种黏土、砂、水和沥青（一种重质黑色稠油）的混合物。

hydraulic and electrically powered shovel *np.* 液压电动挖土机

例：These systems use large hydraulic and electrically powered shovels to dig up tar sands and load them into enormous trucks that can carry up to 320 tons of tar sands per load.

这些方法包括使用大型液压挖土机和电动挖土机挖掘焦油砂，然后装载到巨型卡车上，这种卡车每次可负载多达 320 吨的焦油砂。

bitumen *n.* 天然沥青

例：The bitumen in tar sands cannot be pumped from the ground in its natural state.

油砂中的天然沥青无法以其天然相态从地下泵上来。

strip mining /open pit mining *np.* 露天开采

例：Tar sand deposits are mined, usually using strip mining or open pit techniques, or the oil is extracted by underground heating with additional upgrading.

油砂沉积物可以开采出来，通常使用露天开采技术或地下加热处理技术开采。

in–situation production method/in situ recovery *np.* 现场开采法 / 原地提取

例：In-situation production methods are used on bitumen deposits buried too deep for mining to be economically recovered.

现场开采法适用于埋藏太深、不宜开采，且开采成本高的沥青质沉积物。

extraction plant *np.* 抽提装置

例：After mining, the tar sands are transported to an extraction plant, where a hot water process separates the bitumen from sand, water, and minerals.

焦油砂被挖出后，运至抽提装置，经过热水处理，沥青质便与砂、水以及矿物质分离开来。

light hydrocarbon *np.* 轻烃

例：Because it is so viscous (thick), bitumen requires dilution with <u>light hydrocarbons</u> to make it transportable by pipelines.

由于沥青质具有高黏性，所以需要用轻烃进行稀释，以便可以用管道运输。

synthetic crude oil/syncrude *np.* 合成原油

例：<u>Synthetic crude oil</u>, produced from tar sand (oil sand) bitumen, is also used as feedstocks in some refineries.

合成原油是从焦油砂（油砂）沥青中提取的，在一些炼油厂中也作为原料使用。

upgrading facility *np.* 精加工设施

例：The extracted bitumen is diluted with a special solvent and then sent via pipeline to an <u>upgrading facility</u>, where it is transformed into a wide range of premium low-sulfur and low-viscosity synthetic crude oils.

采出的沥青质要用一种特殊溶剂进行稀释，然后管输至精加工设施，转变为不同类别的优质低硫、低黏度合成原油。

hydrocarbon wetted *adj.* 亲烃的

例：Canadian tar sands are different than U.S. tar sands in that Canadian tar sands are water wetted, while U.S tar sands are <u>hydrocarbon wetted</u>.

加拿大焦油砂与美国的有所不同，前者亲水，后者亲烃。

agitate *v.* 搅拌

例：Hot water is added to the sand, and the resulting slurry is piped to the extraction plant where it is <u>agitated</u>.

焦油砂加高温热水后产生钻井液，钻井液通过管道运至提炼厂，进行搅拌。

spent sand *np.* 废弃油砂

例：After oil extraction, the <u>spent sand</u> and other materials are then returned to the mine, which is eventually reclaimed.

采油结束后，废弃油砂及其他物质都要运回矿场，使矿场恢复再生。

fireflood *n.* 火驱

例：In-situ production techniques include steam injection, solvent injection, and firefloods, in which oxygen is injected and part of the resource burned to provide heat.

现场开采法包括注蒸汽法、注溶剂法和火驱法（注入氧气，燃烧部分油砂提供热量）。火驱是向地层注入氧气，燃烧部分油气提供热量。

natural gas liquid（NGL） *np.* 天然气凝析液

例：Natural gas liquids (NGLs) are the raw, associated gases and liquids that come up along with oil and natural gas from the well.

天然气凝析液是随原油和天然气一起采出井的湿气、伴生气和伴生液。

extra heavy oil *np.* 超重原油

例：Extra heavy oils are extremely viscous, with a consistency ranging from that of heavy molasses to a solid at room temperature.

超重原油非常黏稠，在室温下类似稠糖浆或固体。

oil shale *np.* 油页岩

例：Oil shale is an organic-rich fine-grained sedimentary rock from which liquid hydrocarbons (shale oil) can be extracted.

油页岩是富含有机质的细粒沉积岩，可以从中开采出液态烃（页岩油）。

dry gas well *np.* 干气井

例：US producers have tried to offset the impact of depressed gas prices by shifting their drilling from so-called "dry gas wells", which produce only gas, to "wet wells", which produce a mix of gas and more expensive oil liquids for the petrochemical industry.

鉴于天然气价格低迷，美国生产商不再钻所谓只产气的"干气井"，而是转向钻"湿气井"，湿气井可产出气液混合体，其中液体价格更高，适用于石油化工业。

dense rock formation *np.* 致密岩层

例：Hydraulic fracturing involves injecting a high-pressure mix of water, sand and chemicals to break and hold open dense rock formations, allowing crude, gas and

NGLs to flow to the surface.

水力压裂即高压注入水、砂和化学品混合物，冲进致密岩层并保持其张开，使原油、天然气和天然气凝析液流到地面。

light tight *np.* 轻致密油

例：Light tight oil is analogous in many ways to shale gas, both in terms of its origins—it is oil that has not migrated, or at least not migrated far, from the (shale) source rock.

轻致密油在许多方面和页岩气相似，尤其在来源方面，此类油气在烃源岩中一般不运移，即使运移也不会远离烃源岩。

chemical additive *np.* 化学添加剂

例：A standard single-stage hydraulic fracturing may pump down several hundred cubic meters of water together with proppants and a mixture of various chemical additives.

一段水力压裂一般要注入几百立方米水，以及支撑剂和各种化学添加剂。

produced stream *np.* 产出流体

例：Once the hydraulic fracturing has been completed, some of the fluid injected during the process flows back up the well as part of the produced stream, though typically not all of it—some remains trapped in the treated rock.

水力压裂一旦完成，部分注入的流体会随产出流体一起沿井返排出来，但不会全部排出，一部分仍然会滞留在岩石中。

green completion/reduced emissions completion *np.* 绿色完井/减少排放完井

例：Best practice during this period is to use a "green completion" or "reduced emissions completion", whereby the hydrocarbons are separated from the fracturing fluid (and then sold) and the residual flow-back fluid is collected for processing and recycling or disposal.

这时最好进行"绿色完井"或"减少排放完井"，以便从压裂液中分离出烃来出售，将残余返排压裂液收集起来，处理后再循环使用或废弃。

pour point depressant (PPD) *np.* 降凝剂

例：The current research work embarks on the development of novel nanocomposite pour point depressant (PPD) for waxy crude oils.

目前主要针对含蜡原油开发新型纳米复合降凝剂（PPD）。

well abandonment/abandoned well *np.* 弃井

例：Much of the well abandonment will not take place until production has ceased.

只有停产才会弃井。

greenhouse gas emission *np.* 温室气体排放

例：Greenhouse gas emissions of tar sands crude are 40% higher than that of conventional oil, and the open-pit mining has devastated Alberta's boreal forest.

焦油砂的温室气体排放量比常规石油高 40%，露天采油已经极大地破坏了阿尔伯塔的北方气候带森林。

seismic event *np.* 地震

例：Larger seismic events can be generated when the well or the fractures happen to intersect, and reactivate an existing fault.

当井筒或裂缝恰巧交切时，会激活已有断层，产生较大地震。

Part IV

Petroleum Storage and Transportation
油气储运

Section I

Oil-gas Gathering and Transportation
油气集输

oil–gas gathering and transportation *np.* 油气集输

例：Oil-gas gathering and transportation is the follow-up link to oil production from the stratum.

油气集输是地层采油的后续环节。

gathering and transportation system *np.* 集输系统

例：Oil field gas gathering and transportation system generally includes gas gathering, gas processing; Transportation of dry gas and light hydrocarbon; Sealed transportation of crude oil, stability of crude oil, storage of light hydrocarbon, etc.

油田天然气集输系统一般包括：天然气集输、天然气处理；干气和轻烃的运输；原油的密封运输、原油稳定、轻烃储存等。

gathering pipeline network/field–gathering system *np.* 集输管网

例：Oil & gas gathering pipeline network is a significant part of oil & gas field construction, and the rational construction of pipeline network is directly related to the efficiency and benefits of oil & gas field production.

油气集输管网是油气田建设的重要组成部分，管网的合理建设直接关系到油气田生产效率及效益。

well fluid *np.* 井流

例：Well fluids are often a complex mixture of liquid hydrocarbons, gas, and some impurities.

井流通常是一种液态烃、气体和某些杂质的复杂混合物。

Part IV　Petroleum Storage and Transportation　油气储运

flow line　*np.* 出油管 / 采气管

例：<u>Flow lines</u>, the first link in the transportation chain from producing well to consumer, are used to move produced oil from individual wells to a central point in the field for treating and storage (a tank battery).

出油管是把油从油井输送到用户过程的第一个环节。用出油管把油井产出的油输送到处理中心进行预处理和储存（罐区）。

internally coated pipe　*np.* 内涂层管道

例：Whether or not <u>internally coated pipe</u> is used depends on the corrosion potential of the oil, the expected producing life of the well being served, and other factors.

是否使用内涂层管取决于原油的腐蚀性及油井的生产寿命和其他因素。

intrafield pipeline　*np.* 油气田内管道

例：A flow line is an <u>intrafield pipeline</u>, in the sense that it is used to connect subsea wellheads, manifolds and the platform within a particular development field.

管路是油气田内部的管道，其作用是连接某开发油气田中的海底井口、支管以及特殊的生产平台。

gathering line　*np.* 集输管道

例：The next link in the oil pipeline chain is <u>gathering lines</u>, which are usually short-a couple of hundred meters and with small diameters.

输油过程的第二个环节是集输管道，通常比较短，一般只有几百米，且直径较小。

treatment plant　*np.* 处理厂

例：Gathering pipelines are a group of smaller interconnected pipelines forming complex networks with the purpose of bringing crude oil or natural gas from several nearby wells to a <u>treatment plant</u> or processing facility.

集输管道由一系列小型且相互连接的管道组成，这些管道组成复杂的管网，目的是将附近井口产出的原油或天然气运至处理厂或加工设备处。

transportation pipeline　*np.* 运输管道

例：<u>Transportation pipelines</u> are mainly long pipes with large diameters, moving products (oil, gas, and refined products) between cities, countries and even continents.

运输管道主要由大管径的长管道组成，这些管道在城市、国家甚至大陆间运输产品（石油、气体及成品油）。

export pipeline *np.* 外输管道

例：The pipeline, sometimes referred to as an "export pipeline", is used to bring the resource to shore.

管道有时也称"外输管道"，其作用是将资源运输至陆地上。

distribution pipeline *np.* 分配管道

例：Distribution pipelines are composed of several interconnected pipelines with small diameters, used to take the products to the final consumer.

分配管道由几个相互连接的小直径管道组成，用于将产品输送至终端用户。

feeder line *np.* 集油支线

例：Feeder lines distribute gas to homes and businesses downstream.

集油支线将气体分输到下游的家庭及商业用户。

central tank farm *np.* 中央油库

例：Crude oil gathering station is used for gathering, treating or storing crude oil and includes central tank farm, oil collecting station, gas compressor station and well head installation.

原油集油站用于收集、处理和储存原油，包括中央油库、集油站、气体压缩机站和井口等装置。

jumper *n.* 跨接管

例：Indeed, it is pivotal that oil and gas are transported from the reservoir to a refinery through a process that involves the fluids flowing from the wellhead through a jumper to a manifold.

实际上，油气从油气藏运至炼油厂需要这样的过程，即井口流出的流体通过跨接管进入地面管汇。

flow rate *np.* 流量

例：The size of the opening in the choke determines the flow rate, because the pressure upstream is determined primarily by the well FTP.

油嘴的开孔直径决定了油流流量，因为油流的上游压力主要取决于油井的自喷油压。

gas transmission pipe line/gas trunk line *np.* 输气干线

例：Gas transmission pipe lines are made of steel pipe and buried below ground surface, which is externally coated to protect against corrosion.

输气干线都使用钢管，并埋于地下，外部涂有防腐层。

gas-gathering pipeline *np.* 集气管线

例：The purpose of gas-gathering pipelines and gas transmission lines is similar to that of crude-gathering and crude trunk lines, respectively, but operating conditions and equipment are quite different.

集气管线和输气干线的作用分别与集油管线和原油干线的作用相似，但管道的操作条件及使用的设备与原油管道不同。

compressor *n.* 压缩机

例：In general, gas pipelines operate at high pressures than crude lines; gas is moved through a gas pipeline by compressors rather than by pumps.

气体管道运行压力一般比原油管道高，因此气体管道不用泵，而用压缩机使气体运行。

pressure-reducing valve *np.* 减压阀

例：Many gas wells produce at such high pressures that pressure must be reduced at the wellhead by a choke or pressure-reducing valve before the gas enters the flow line.

多数气井井口压力很高，必须在井口使用节流阀或减压阀降低气体压力后，气体才能进入采气管。

pressure control valve *np.* 压力控制阀

例：The pressure downstream is determined primarily by the pressure control valve on the first separator in the process.

下游压力主要取决于第一个分离器上的压力控制阀。

positive choke *np.* 固定油嘴

例：For high-pressure wells it is desirable to have a <u>positive choke</u> in series with an adjustable choke.

高压井口上，最好安装一个固定油嘴，并串联一个可调油嘴。

adjustable choke *np.* 可调油嘴

例：The positive choke takes over and keeps the production rate within limits should the <u>adjustable choke</u> fail.

若可调油嘴失控，则固定油嘴可取而代之，并将采出的油流量控制在规定范围内。

automatic shutdown valve *np.* 紧急自动切断阀

例：On offshore facilities and other high-risk situations, an <u>automatic shutdown valve</u> should be installed on the wellhead.

在海上采油设备中及其他有风险情况下，采油井口上必须装有紧急自动切断阀。

test production system *np.* 采油测试系统

例：Whenever flows from two or more wells are commingled in a central facility, it is necessary to install a manifold to allow flow from any one well to be produced into any of the bulk or <u>test production systems</u>.

只要两口以上的井流汇入一个处理中心，都必须安装汇管，以便油井所采油流都能进入各容器或采油测试系统。

liquid recovery *np.* 液量采收

例：As more stages are added to the process there is less and less incremental <u>liquid recovery</u>.

原油处理过程中分离级数越多，所能采收的液量就越少。

primary separator *n.* 一级分离器

例：By using a manifold arrangement and different <u>primary separator</u> operating pressures, there is not only the benefit of stage separation of high-pressure liquids, but also conservation of reservoir energy.

配置好管汇及一级分离器的各种运行压力，不仅有利于高压井液的多级分离，也会保护油层能量。

central field facility *np.* 油田中心处理设施

例：The pressure that forces oil in a flowing well to flow to the surface is usually sufficient to move the oil on to the central field facility.

如果自喷井压力足以把油从井底运到地面，通常也足以把油输送到油田中心处理设施。

storage terminal *np.* 转运仓库

例：From large central storage facilities, oil is moved through large-diameter, long-distance trunk lines to refineries or to other storage terminals.

原油通过大直径、长距离原油干线，从大型中央储存设施输送到炼油厂或其他储存终端。

bottom hole pump *np.* 井底泵

例：A bottom hole pump must be used in some wells to lift the oil to the surface.

有些油井必须使用井底泵将油泵到地面。

flow assurance *np.* 流动安全

例：The goal of any pipeline designer is to secure "flow assurance", i.e., the transmission system must operate in a safe, efficient, and reliable manner throughout the design life.

所有管线设计者的目标都是确保流动安全，也就是说，集输系统在设计的使用期限内必须安全、有效、平稳地运转。

lease *n.* 矿区 / 油田

例：A typical tank battery contains a separator to separate oil, gas, and water; a fired heater to break water-oil emulsions to promote complete removal of water from the oil; and tanks for storing the oil until it can be shipped from the lease.

罐区一般设有用于油、气、水分离的分离器，对乳化液进行油、水分离的加热炉和运离矿区前储存油品的储罐。

heated storage *np.* 加热储存

例：Heated storage may be required if the oil is too viscous at low temperatures to be pumped from lease storage.

如果低温时原油黏度过大，用泵外输时，就需要对罐内原油进行加热。

metering equipment/facility *np.* 计量设备

例：Metering equipment is also included to measure the volume of oil leaving the lease.

计量外输油量还要用到计量设备。

positive volume metering *np.* 容积式流量计

例：Positive volume metering involves the filling of a predetermined volume, the automatic discharge of that volume by liquid level-actuated valves, and the recording of the discharge by some type of counter.

容积式流量计的使用涉及预定容积的填充、液位驱动阀门自动排出该容积、某种类型计数器记录排出量。

GLR *abbr.&np.* 气液比

例：GLR is the abbreviation for gas/liquid ratio, the ratio of produced gas to produced liquids (oil and water).

气液比是气/液比的缩写，指采出气体与采出液体（油和水）的比值。

GOR *abbr.&np.* 油气比

例：GOR is the abbreviation for gas/oil ratio, the ratio of produced gas to produced oil.

油气比是气/油比的缩写，即产气与产油的比值。

WOR (water oil ratio) *abbr.&np.* 水油比

例：The ratio of produced water to produced oil is abbreviated as WOR.

采出水与采出油之比，简称水油比。

gathering system throughput *np.* 集输系统的输送能力

例：Gathering system throughput obviously varies widely, depending on the number of field storage tanks served and the producing capacity of wells in each field.

集输系统的输送能力显然差别较大，取决于每个油田储罐的数量和油井生产能力。

utility *n.* 公用事业公司

例：Through distribution networks of small pipelines and metering facilities, utilities distribute natural gas to commercial, residential, and industrial users.

通过小直径配气管网和计量设备，公用事业公司把天然气配送给工商业用户和居民。

metering separator *np.* 计量分离器

例：Metering separators accumulate the separated oil, or oil and water, isolate the liquid phases in calibrated chambers, and periodically discharge the measured volumes into separate outlets.

计量分离器积聚分离的油或油和水，在校准室中分离液相，并定期将测量的体积排放到单独的出口中。

Venturi tube *np.* 文丘里管 / 文氏管

例：Both flow nozzles and Venturi tubes can handle about 60% more flow than an orifice plate under the same conditions, and therefore are often used to handle higher velocity flows.

喷嘴和文氏管能处理的流体流量比孔板大 60％，所以经常用于处理高速流体。

flow coupling *np.* 流量联轴器

例：A flow coupling is a short, heavy-duty tubing joint run above and below tubing restrictions (safety valves, chokes, etc.) that minimize abrasive effects of turbulent flow caused by the restrictions.

流量联轴器是一种短的、耐用的油管接头，在油管上、下限位（安全阀、扼流圈等）运行，最大限度地减小由限位引起的湍流 / 紊流磨损效应。

single-stage process *np.* 单级分离

例：In a simple single-stage process, the fluids are flashed in an initial separator and then the liquids from that separator are flashed again at the stock tank.

在单级分离中，流体在初级分离器中闪蒸后，出来的液体再次在存液罐中闪蒸。

floating production storage and offloading (FPSO) *np.* 浮式生产储存卸货装置

例：Floating production storage and offloading (FPSO) refers to a floating vessel located near an offshore oil field, where oil is processed and stored until it can be transferred to a tanker for transporting and additional refining.

浮式生产储存卸货装置（FPSO）是指海上油田附近的浮式卸储船舶，石油在此处经过加工和储存，直至可转移至槽车运输并进一步提炼。

skid-mounted equipment *np.* 撬装设备

例： We offer stand-alone and skid-mounted equipment, both for onshore facilities and offshore platforms and topsides on FPSO.

独立撬装设备，既适用于陆上设施，也适用于海上平台和浮式生产储油船顶部。

processing plant *np.* 加工厂

例： Oil and gas gathering pipeline move gas and oil from drilling sites to processing plants, refineries.

通过油气集输管道，把天然气和石油从钻井现场输送到加工厂、炼油厂。

Oil and gas pipe gallery *np.* 油田管廊带

例： Oil and gas pipe gallery is an important channel for oil and gas gathering and transportation and material transmission in chemical industry park.

油田管廊带是油气集输和化工园区物料传输的重要通道。

well tube *np.* 配水管

例： However, with the reservoir pressure decreasing and water production gradually increasing, field test show it could be used to enhance entrainment capability of well tube.

但随着油藏压力降低，产水量逐渐增加，现场试验表明，该技术可以提高配水管的携液能力。

loading arm *np.* 装油鹤管

例： Loading arm is one important fluid transfer equipment nowadays in the petrochemical industry, mainly used for transferring the fluid such as petroleum, LNG, LPG, chemicals, gas etc, between pipe system and the tankers.

装油鹤管是当前石油化工行业中重要的流体输送设备之一，主要用于石油、液化天然气、液化石油气、化学品、天然气等流体在管道系统和油轮之间的输送。

gas gathering station *np.* 集气站

例： Paramount Limited designs and installs gas gathering stations to separate the liquid from the gases that are pumped from oil producing wells.

派拉蒙有限公司（Paramount Limited）设计并安装集气站，以分离从采油井泵出的液体和气体。

lease automatic custody transfer (LACT) *np.* 矿场自动接收、取样、计量、传输系统（LACT）

例： Figure 2 shows schematically the elements of a typical lease automatic custody transfer (LACT) unit.

表 2 显示了常规的矿场自动接收、取样、计量、传输系统中的组件及流程。

processing facility *np.* 处理设施

例： Some natural gas gathering systems include a processing facility, which performs such functions as removing impurities like water, carbon dioxide or sulfur that might corrode a pipeline, or inert gases, such as helium, that would reduce the energy value of the gas.

某些集气系统有处理设施，可除去腐蚀管道的水、二氧化碳或硫等杂质，或去除可降低天然气热值的氦气等惰性气体。

center processing facilities (CPF) *np.* 中央处理站

例： The arrangement of gas gathering stations (GGS) and central processing facilities (CPF) in the natural gas gathering pipeline network really counts.

集气站（GGS）和中央处理站（CPF）的排布对天然气集输管网至关重要。

multiphase flow transmission system *np.* 多相流集输系统

例： In thermal-hydraulic design of multiphase flow transmission systems, the system designer is faced with several challenges associated with multiphase flow, which can significantly change design requirements.

对多相流集输系统进行热力和水力设计时，设计者面临着诸多难题，这些难题都与多相流动有关，多相流大幅提高了设计要求。

Section II

Oil and Gas Storage
油气储存

storage facility *np.* 储存设施

例：Oil and gas separation, treating, and storage facilities are commonly referred to as a tank battery.

油气分离、处理和储存设施统称为罐群。

delivery terminal *np.* 交货终点站

例：Tank storage facilities exist at the origin terminal of a pipeline, at the delivery terminal of a pipeline and at refineries served by these pipelines.

在管道的首发站、交货终点站及这些管道涉及的炼油厂都有油罐储存设施。

tank battery *np.* 选油站 / 油罐区

例：Gunbarrel tanks are an excellent choice for tank batteries when atmospheric pressure, gravitational force, and retention time are sufficient for the adequate separation of oil, water, and gas.

当大气压力、重力和停留时间足以使油、水和气体充分分离时，油罐区使用油水分离沉降罐是最佳选择。

oil tank *np.* 油罐

例：Oil tanks are ideal containers used to store crude oil or other kind of oil, they are widely used in refinery, oil field, oil tank farm, or other industrial production.

油罐是储存原油和其他石油的理想容器，广泛应用于炼油厂、油田、油库和其他工业生产中。

petroleum storage tank *np.* 石油储罐

例：A petroleum storage tank is designed to store a range of industrial liquids and oil,

Part IV Petroleum Storage and Transportation 油气储运

particularly the crude variant.

石油储罐用于储存一系列工业液体和石油，特别是原油。

aboveground storage tanks (ASTs) *np.* 地面罐

例：Aboveground storage tanks (ASTs) serve the purpose well fulfilling storage needs of the industry.

地面罐可以较好地满足油气行业的存储需求。

underground storage tanks (USTs) *np.* 地下罐

例：Approximately 542,000 underground storage tanks (USTs) nationwide store petroleum or hazardous substances.

全国约有 54.2 万地下罐存储着石油和有害物质。

semi–underground storage tank *np.* 半地下储罐

例：Semi-underground storage tank refers to an oil tank whose burial depth exceeds half of the tank height, and the highest oil level in the oil tank is not higher than the lowest elevation of the adjacent area by 2m.

罐底埋入地下深度不小于罐高的一半，且罐内液面不高于附近地面最低标高 2m 者为半地下储罐。

elevated storage tank *np.* 高架储罐

例：Elevated storage tanks are a great choice for storage of various oils, antifreeze, diesel exhaust fluid (DEF), and more.

高架储罐存储功能强大，可以存储多种油类、防冻液、柴油尾气处理液（DEF）等。

low pressure gas tank *np.* 低压储气罐

例：Low pressure gas tanks are suitable for the storage of liquids which are too volatile for atmospheric storage.

低压储气罐适用于储存易挥发、不适于常压储存的液体。

atmospheric storage tank *np.* 常压储气罐

例：An atmospheric storage tank stores products at the same pressure as the atmosphere at the location of the tank.

常压储气罐的压力与储罐所在位置的大气压力相同。

high pressure gas tank *np.* 高压储气罐

例：The high pressure gas tank is filled with high-pressure air with pressure higher than working pressure of the dredge pump and supplies an air inlet pipe and an air outlet tube of the underwater pneumatic dredge pump with high-pressure air.

高压储气罐内充满压力高于疏浚泵工作压力的高压空气，为水下气动疏浚泵的进气管和出气管供给高压空气。

fixed-roof tank *np.* 固定顶储罐

例：A "pressure" fixed-roof tank designed for a maximum internal working pressure of 203 kgf/m^2 will approximately halve, on average, the breathing losses from gasoline storage in temperate climates, compared with similar storage in an open vented tank.

根据最大内部工作压力 203 kgf/m^2 设计的"压力"固定顶储罐，在温带气候下储存汽油，与通大气罐类似储存条件相比，呼吸损耗的平均值约减少一半。

pressure roof tank *np.* 压力顶罐

例：Pressure roof tanks designed for a maximum pressure of 546 kgf/m^2 are limited to a maximum of 19.5 m diameter.

设计最大压力为 546 kgf/m^2 的压力顶罐，最大直径限制为 19.5m。

dome-roof storage tank *np.* 拱顶罐

例：The normal operation of dome-roof storage tank includes receiving oil, storing oil, sending oil and measuring oil.

拱顶罐的正常运行流程包括收油、储油、送油和测油。

floating roof tank *np.* 浮顶油罐

例：Floating roof tanks are used in the oil and gas industry to store large quantities of highly flammable or volatile chemicals like crude oil.

浮顶油罐用于石油和天然气行业中，用于储存大量像原油一样的高度易燃或挥发性化学品。

static electric charge *np.* 静电荷

例：A secondary advantage is an increase in operational safety brought about by the absence of a vapor space above the liquid (since the roof rests on the liquid) and the immediate dissipation by the floating roof of any static electric charges on the liquid.

第二个优点是增加了操作的安全性。因顶浮在液体上，液体上方无蒸气空间，即使产生液体静电荷也可通过浮顶立即消散。

vapor-tight seal np. 密封不漏蒸气

例：The steel roof floats on the liquid and moves up and down as oil is pumped into or out of the tank, a significant feature of the tank design being the vapor-tight seal between the rigid periphery of the floating roof and the tank shell.

钢顶浮在液面上，并随油的泵入、泵出而上下移动，这种设计特点是浮顶刚性边缘与罐壁间密封极好不漏蒸气。

volatile liquid np. 挥发性液体

例：The floating-roof tank has met the need for economical bulk storage of volatile liquids with a high degree of safety.

浮顶油罐能经济储存大量挥发性液体，安全性高。

external floating roof tank np. 外浮顶罐

例：In oil and gas industry, the external floating roof tanks are the latest developments and are known for their cost-effectiveness, efficacy, and safety.

在石油和天然气行业中，最新研发的外浮顶罐以低成本、高效性和安全性著称。

floating roof np. 浮盘

例：The floating roofs can be either external (i.e., with open top tanks) or internal (i.e., inside fixed roof tanks).

浮盘可以安装在油罐内部（开顶罐），也可以安装在油罐外部（内固定顶油罐）。

internal floating roof storage tank np. 内浮顶罐

例：Compared with external floating roof tank, internal floating roof storage tank has a fixed roof, which is beneficial to improve the storage condition of oil products.

与外浮顶罐相比，内浮顶罐有固定顶盖，更利于改善油品的储存条件。

pan-type roof *np.* 盘形顶

例：In its early form, the floating roof consisted of a single deck with a vertical rim at the periphery, which was the <u>pan-type roof</u>.

早期的浮顶是由周边装有垂直边缘的单盘构成的，这种浮顶是盘形顶。

annular pontoon *np.* 环形浮舱

例：The next step was to give the roof greater buoyancy by fitting an <u>annular pontoon</u> around the periphery, the pontoon encircling a single center deck.

下一步是通过在顶的周边安装环形浮舱以增加顶的浮力，浮舱在中央单盘的外围。

pontoon roof *np.* 浮舱式顶

例：<u>Pontoon roofs</u> are suitable for those volatile liquids with R. V. P. up to 0.84 kgf/cm^2.

浮舱式顶适合于那些雷德蒸气压不超过 0.84kgf/cm^2 的挥发性液体。

double-deck construction *np.* 双盘结构

例：Further development of the floating roof was in the direction of maximum buoyancy and stability provided by a <u>double-deck construction</u>.

浮顶进一步发展的方向是双盘结构，以提供最大浮力和稳定性。

vapor pressure *np.* 蒸气压

例：The pressure immediately below the floating roof is roughly atmospheric and, for this reason, floating-roof tanks are not recommended for the storage of liquids whose <u>vapor pressure</u> exceeds atmospheric pressure at the storage temperature.

紧靠浮顶下面的压力大致等于大气压，因此，浮顶罐不推荐用于在储存温度下蒸气压超过大气压的液体。

rigid floating roof *np.* 刚性浮顶

例：The <u>rigid floating roof</u>, whether it be of the pan, pontoon, or double-deck type, must be able to move freely up and down with the liquid.

刚性浮顶，无论是盘型、浮舱型还是双盘型，都必须能随液体上下自由运动。

dump tank *np.* 卸油罐

例：The researchers evaluated each <u>dump tank</u> through two water cycles, meaning that if a packing house changed the water every three days.

研究人员通过两个水循环来评估每个卸油罐，这意味着包装厂需要每三天更换一次水。

slop tank *np.* 污油罐

例：To overcome this problem the refinery should change the oil analyzing program by analyzing the separated oil in the <u>slop tanks</u> in the first group.

为解决这一问题，炼油厂应改变油品分析程序，对第一组污油罐内分离出的油进行分析。

gunbarrel tank *np.* 沉降罐

例：At onshore locations the oil may be treated in a big "<u>gunbarrel</u> (or settling) <u>tank</u>".

陆地采油设备中，原油是在大型"沉降罐"中进行处理的。

surge tank *np.* 缓冲罐

例：Flow from the treater or gun barrel goes to a <u>surge tank</u> from which it either flows into an oil barge or truck or is pumped into a pipeline.

油流经过处理器或沉降罐之后进入缓冲罐，再由缓冲罐流入油驳或油罐车，或泵入管道。

liquefied natural gas（LNG）storage tank/LNG container *np.* 液化天然气储罐

例：<u>liquefied natural gas (LNG) storage tanks</u> have double containers, where the inner contains LNG and the outer container contains insulation materials.

液化天然气储罐采用双层容器，内层容器装液化天然气，外层容器装隔热材料。

LNG cryogenic storage tank *np.* 液化天然气低温储罐

例：The <u>LNG cryogenic storage tank</u> is a typical vacuum powder insulated cryogenic pressure vessel with a double-layered cylinder structure.

液化天然气低温储罐是典型的双层圆筒结构真空粉末绝热低温压力容器。

gas liquefaction or vaporization plant *np.* 天然气液化厂、气化厂

例：Short pipelines, however, are in operation in association with <u>gas liquefaction or vaporization plants</u> and terminals for loading and unloading of LNG tankers.

然而，与天然气液化厂、气化厂和海轮装船、卸船的终点相连的仍是短距离管道。

cryogenic storage tank *np.* 深冷储罐

例：Large cryogenic storage tanks are needed to store the liquefied natural gas; typically these may be 70 m in diameter, 45 m high, and hold over 100,000 m³ of liquefied natural gas.

天然气储存需要大型深冷储罐，一般储罐直径为 70 米、高为 45 米，能储存 100000 立方米的液化气。

refrigerated tanker *np.* 低温冷藏罐

例：The current largest specially built refrigerated tankers can carry 135,000 m³ of liquefied natural gas, equivalent to 2.86 billion scf of gas, but are very expensive.

目前，世界上最大的低温冷藏罐能容纳 13.5 万立方米的液化气，相当于 28.6×10^8 标准立方英尺的天然气，但价格很高。

spherical tank *np.* 球罐

例：Spherical tank as a petro-chemical industry, one of the most commonly used equipment, the storage material with a flammable, explosive, toxic, corrosive, and other characteristics, has been related to the society constitutes a major public safety hazard.

球罐是石油化工行业最常用的设备之一，存储具有易燃、易爆、剧毒、强腐蚀等，以及对社会构成重大公共安全隐患的物质。

stock tank *np.* 库存罐

例：This liquid contains some light components that vaporize in the stock tank downstream of the separator.

该液流中含有一些轻质组分，轻质组分流入分离器下游的库存罐内，并逐渐蒸发。

bolted steel tank *np.* 螺接库存罐

例：Bolted steel tanks are normally 500 barrels or larger and are assembled on location.

螺接库存罐的容量通常为 500 桶或更大一些并在现场安装。

welded steel tank *np.* 焊接钢罐

例：Welded steel tanks range in size from 90 barrels to several thousand barrels. Welded tanks up to 400 barrels in capacity are shop-welded and are transported complete unit to the tank battery site.

焊接钢罐的容量从 90 桶至几千桶不等，低于 400 桶的焊接钢罐在工厂焊接，然后整体运往罐群安装地点。

hard container *np.* 硬质容器

例：Compressed natural gas (CNG) is stored in hard containers at a pressure of 20–25 MPa, usually in cylindrical or spherical shapes.

压缩天然气通常存储在圆柱形或球形硬质容器中，压力为 20 ~ 25 兆帕。

tank bottom plate *np.* 罐底板

例：The external corrosion mainly covers the corrosion on tank top head, tank bottom plates and wall plates.

外腐蚀主要包括罐顶、罐底板和外壁的腐蚀。

breather valve *np.* 呼吸阀

例：Mechanical breather valve is an important accessory to protect the safety of the oil tank.

机械呼吸阀是保障油箱安全的重要配件。

strapping table *np.* 容积表

例：Strapping table is a tabular record of tank volume versus height.

容积表用来记录储罐容积与高度。

flashboard valve *np.* 闸板阀

例：When in work, the control system automatically opens or closes the gate of the electric flashboard valve according to the signals sent by the material level meters.

工作中，控制系统会根据料位仪传来的信号自动开启或关闭电动闸板阀闸口。

globe valve *np.* 截止阀

例：Globe valves are linear motion closing-down valves used to start, stop or regulate the flow.

截止阀是线性驱动型关闭阀，用于开、关或调节流量。

check valve *np.* 单向阀

例：Check valves are nonreturn valves that open with fluid movement and pressure, and close to prevent backflow of the pressure to upstream equipment such as pumps and compressors.

单向阀是止回阀，随流体流动和压力开启，随后关闭以防止压力回流到上游压缩机或泵等设备中。

plug valve *np.* 旋塞阀

例：Plug valves are valves with cylindrical or conically tapered "plugs" which can be rotated inside the valve body to control flow through the valve.

旋塞阀是带有圆柱形或锥形"旋塞"的阀门，这些"旋塞"在阀体内可以旋转以控制流经阀门的流量。

ball valve *np.* 球阀

例：A ball valve is a flow control device which uses a hollow, perforated and pivoting ball to control liquid flowing through it.

球阀是流量控制装置，用一个中空、穿孔和旋转的球控制液体流量。

drain valve *np.* 疏水阀

例：A drain valve is a mechanical device used to release excess or unwanted quantities of liquid or gas from a storage tank, vessel or container.

疏水阀是机械装置，用于从储罐、船舶或容器中释放多余的液体或气体。

oil dipstick/oil dip rod *np.* 量油尺

例：The engine oil dipstick is a flexible metal rod that slides down the oil dipstick tube into the oil pan to measure the oil level.

发动机量油尺是一根灵活的金属杆，可以沿着机油尺管滑入油箱，测量机油液位。

water gauge *np.* 量水尺

例：Water gauge is an instrument to measure or find the depth or quantity of water or to indicate the height of its surface especially in a steam boiler.

量水尺是测量水深、水量或表面高度的工具，尤其是用于蒸汽锅炉。

tank gauging *np.* 油罐计量

例：Offshore platforms do not store crude oil and consequently, <u>tank gauging</u> is not applicable on offshore platforms.

海上平台不储存原油，因而不用油罐计量。

level measurement/gauging *np.* 液位计量

例：<u>Level measurement</u> features predominantly in crude oil and petroleum product transportation systems.

液位计量在原油和石油产品输送系统中起着重要作用。

level transmitter *np.* 液位传送器

例：Level gauging (sight glasses), level switches, <u>level transmitters</u> and controllers are installed for the monitoring and control of crude oil in production.

在平台上装有液位计量器（观察玻璃管）、液位开关、液位传送器和控制器来监控原油生产。

level switch *np.* 液位开关

例：<u>Level switch</u> is mainly composed of reed switch and float.

液位开关主要由磁簧开关和浮子组成。

inventory monitoring *np.* 库存油品监测

例：Level gauging, in addition to being critical for custody transfer accounting, is also very critical for <u>inventory monitoring</u> and control, and for pipeline scheduling and dispatching.

液位计量除对输油监测结账（交接计量）很重要，对于库存油品监测和控制、管输计划和调度也很重要。

remote automatic tank gauging system *np.* 遥控自动油罐计量系统

例：Many level gauging systems are based on <u>remote automatic tank gauging systems</u>.

许多液位计量装置以遥控自动油罐计量系统为主。

float–type level gauge *np.* 浮子液位计

例：Remote automatic tank gauging systems are based on a <u>float-type level gauge</u> as the basic measurement device.

遥控自动油罐计量系统以浮子液位计作为基本计量装置。

column of liquid *np.* 液柱

例：Other level measurements are based on the principle of measuring the pressure exerted by a column of liquid in the tank.

其他液位计量是以测量罐中液柱产生的压力为原理。

capacitance *n.* 电容

例：Other types of level measurement are accomplished by ultrasonics, radiation, microwave, capacitance, etc.

其他形式的液位计量是靠超声波、无线电波、微波、电容等来完成的。

level stick/level tape *np.* 液位杆/液位尺

例：Another level measurement device in common use is the level stick or level tape, which is manually lowered into the tank.

还有一种普遍使用的液位计量工具，是人工放到罐内的液位杆或液位尺。

custody transfer accounting *np.* 交接计量

例：This level device is also the basis for custody transfer accounting for crude oil and petroleum products tankage all over the world.

这种液位检尺工具也是全世界原油和石油产品库存交接计量的基准。

automatic level gauge *np.* 自动液位计

例：Manual methods are liable to operational procedural errors and also liable to incorrect data recording, whereas, automatic level gauges have to be calibrated and maintained in calibration.

人工液位计量方法易产生操作程序性误差，也易发生错误的数据记录，而自动液位计则需要在分度时进行标定和维护。

level detection *np.* 液位探测

例：The basic principles used for level measurement are also used for level detection.

液位计量的基本原理也用于液位探测中。

high level alarm *np.* 高液位报警器

例：<u>High level alarms</u> will alert the operator that a particular tank is becoming full and product delivery must be diverted to another tank.

高液位报警器会提示操作员油罐将要装满，所输油品必须切换至另一个油罐内。

tank suction outlet line *np.* 油罐出入管

例：Continued operations would cause the product level to decrease to the <u>tank suction outlet line</u>.

继续操作会使液位降到油罐出入管处。

flashing *n.* 闪蒸

例：This would draw air and vapors into the product which can cause <u>flashing</u> and inaccurate metering.

这会使空气和油蒸气进入油品，从而导致闪蒸和计量误差。

sump pump *np.* 储罐泵

例：Level switches are also employed at pump stations on the sump tank to control the <u>sump pump</u> and to provide a warning when the sump tank is completely full.

在泵站的储罐上也采用液位开关来控制储罐泵，并在储罐完全装满时发出警告。

float *n.* 浮子

例：The basic principle of a tank gauging system is the measurement of the movement of a <u>float</u> that moves, up or down, with the movement of the liquid in the tank.

油罐计量系统的基本原理是测量浮子随罐内液位变化而产生的上、下位移。

float–type tank gauge *np.* 浮子型油罐液位计

例：<u>Float-type tank gauges</u> can be installed on fixed roof, cone roof, lifter roof and floating roof tanks.

浮子型油罐液位计可以安装在固定顶、锥顶、升降顶和浮顶等各类罐上。

central control unit *np.* 中心控制台

例：Tank temperature may also be digitally transmitted to a <u>central control unit</u>.

油罐温度也可以数字方式传送到中心控制台。

spot or average RTD probe *np.* 点状或平均电阻式温度检测器

例：This is achieved by installation of a spot or average RTD probe on the tank, analogue-to-digital conversion at the gauge head, and digital data transmission on receipt of tank address code for level and temperature data.

可在罐上安装一个点状或平均电阻式温度检测器，在表头内进行模拟-数字转换，并在收到油罐的地址码后传送液位和温度数据。

drainage pump station *np.* 排涝泵站

例：Most of the drainage pump stations in the oil field were built in the 1980s.

油田矿区的排涝泵站大多建于20世纪80年代。

transfer pump *np.* 输油泵

例：Fuel and oil transfer pumps move oil, fuel, lubricants, and other substances from one container to another.

输油泵的作用是将石油、燃料、润滑剂和其他物质从一个容器输送到另一个容器。

gear pump *np.* 齿轮泵

例：High pressure gear pumps have enjoyed a long history in the oil and gas industry.

高压齿轮泵在石油和天然气行业中历史悠久。

collection space *np.* 围堰区

例：The storage tanks are placed in collection spaces which are made impermeable to oil using clay layers, plastic tilts, or concrete lining.

储罐集中置放于由黏土层、塑料罩隔离层或者混凝土衬层建造的不渗透围堰区内。

storage onboard *np.* 船上储存

例：The gas has to be dried, compressed, and chilled for storage onboard.

气体必须经过干燥、压缩和冷却才能船上储存。

hydrocarbon emission *np.* 烃类泄漏

例：Most hydrocarbon emissions in processing occur in the storage areas, ie, tank farms for crude, feedstocks, intermediate, and final products.

生产过程中发生的烃类泄漏大多发生在储存区，例如原油罐区、原料储库、中间品和终端产品储库等。

emission-free tank farm *np.* 零泄漏油库

例：More recent developments are <u>emission-free tank farms</u>, where several fixed roof tanks fitted with internal floating roofs breathe into a closed system at one common gasometer, which normally absorbs all changes in the tank level.

最近在零泄漏油库方面有了较多发展，即几个装有内浮顶的固定顶储罐联入一个闭路系统，共用一个气量计，而气量计通常可以平衡所有油罐的液位。

tank farm *np.* 油库

例：Gathering lines transport oil from field-processing and storage facilities to a large storage tank or <u>tank farm</u> where it is accumulated for pumping into the long-distance crude trunk line.

集油管道把油从油田处理和储存设施输到大型油库，然后再泵入原油长输干线。

ignition temperature *np.* 燃点

例：Water fog is useful in shutting off air from burning oil surface and cooling it below <u>ignition temperature</u>.

水雾有助于隔绝燃油与空气，并使油冷却到燃点以下。

high flash oil *np.* 高闪点油

例：Water fog is effective on viscous oils or <u>high flash oils</u> in areas that are within the range of the spray.

水雾仅对水雾覆盖范围内的黏性油或高闪点油有效。

foam extinguishment system *np.* 泡沫灭火系统

例：There are three main types of <u>foam extinguishment systems</u>.

泡沫灭火系统主要有三种形式。

fixed system *np.* 固定系统

例：The <u>fixed systems</u> have been used where large storage tanks are in use such as large marine terminals or bulk storage plants.

固定系统一般用在有大型储罐的地方，如大型海上转输站或大型储油库。

semi-fixed system *np.* 半固定式系统

例：The semi-fixed system is the same as the fixed system with respect to foam discharge devices except that there is no central foam solution production and storage.

半固定式系统的泡沫排放装置与固定式系统一样，但没有集中的泡沫溶液生成和储存部分。

foam-proportioning truck *np.* 泡沫比例车

例：Instead, mobile foam-proportioning trucks or trailer mounted units are used to produce the foam solution.

相反，使用移动式泡沫比例车或拖车装置来生产泡沫溶液。

back-up truck *np.* 辅助卡车

例：A back-up truck is usually used for resupplying the units with foam concentrate.

辅助卡车通常用于为装置补充泡沫浓缩液。

portable system *np.* 移动式系统

例：Portable systems may be a combination of mobile and hand transportable equipment.

移动式系统可以由车载和人工移动设备组合而成。

foam discharge device *np.* 泡沫排放装置

例：The foam discharge devices may be foam towers, monitor nozzles or foam nozzles.

泡沫排放装置可以是泡沫炮、水枪喷管和泡沫喷管。

spill fire *np.* 流散火灾

例：Portable systems is especially suited for spill fires.

移动式系统特别适用于扑灭流散火灾。

Section III

Oil, Gas and Water Separation
油气水分离

unprocessed well fluids *np.* 未经处理的井筒流体

例：Multiphase transport has become increasingly important for developing marginal fields, where the trend is to economically transport <u>unprocessed well fluids</u> via existing infrastructures, maximizing the rate of return and minimizing both capital expenditure (CAPEX) and operational expenditure (OPEX).

混相输送对海上油田开发越来越重要。如何利用现有基础设施经济有效地输送未经处理的井筒流体，实现收益最大化和固定资产投资（CAPEX）及操作成本（OPEX）最小化是发展趋势。

multicomponent nature *np.* 多组分特点

例：Because of the <u>multicomponent nature</u> of the produced fluid, the higher the pressure at which the initial separation occurs, the more liquid will be obtained in the separator.

由于所采油流具有多组分的特点，因此，在初次分离过程中，油流压力越高，分离器中所能获得的液流就越多。

two–phase gas–liquid flow *np.* 气液两相流

例：The main difference between <u>two-phase gas-liquid flows</u> and three-phase gas-liquid-liquid flows is the behavior of the liquid phases, where in three-phase systems the presence of two liquids gives rise to a rich variety of flow patterns.

气液两相流和气液液三相流的主要区别在于液相的流动状态不同，由于三相流动系统中的液相存在两种液体，这样一来就增加了流动形态多样性。

gas-liquid separator *np.* 气液分离器

例：A conventional separator divides the produced fluid stream into oil and gas, or liquid and gas, and is known as a gas-oil separator or gas-liquid separator.

传统的气液分离器将产出的流体分成油和气，或液和气，称为气油分离器或气液分离器。

multiphase flow *np.* 多相流

例：Multiphase flow in annular conduits might take place in a variety of operational conditions in the production of an oil or gas well.

在油气井生产过程的各种工况下，环形管道中都可能产生多相流。

high-API-gravity crude *np.* 高 API 重度原油

例：Early treatment may be a lesser factor in treating low-viscosity, high-API-gravity crudes.

在处理低黏度、高 API 重度原油时，早期处理作用不大。

natural gasoline *np.* 天然汽油

例：The higher molecular weight hydrocarbons product is commonly referred to as natural gasoline.

天然气中的重烃产物一般称为天然汽油。

light gasoline *np.* 轻质汽油

例：Natural gas can also contain a small proportion of C_{5+} hydrocarbons. When separated, this fraction is a light gasoline.

天然气也可能含有少量的 5 个碳原子以上的碳氢化合物，分离后这些重组分可作为轻质汽油。

immiscible liquids *np.* 不混溶液体

例：For an emulsion to exist there must be two mutually immiscible liquids, an emulsifying agent, and sufficient agitation to disperse the discontinuous phase into the continuous phase.

乳状液形成必须有两种不混溶液体和一种乳化剂，并要充分搅动，使非连续相分散为连续相。

free water *np.* 游离水

例：<u>Free water</u> exists at the same time with oil flow in the process of oil and gas gathering and transportation, but does not emulsify or blend with oil, and maintains a clear interface, which can be separated and discharged at any time.

游离水在油气集输过程中与油流同时存在，但不与油乳化、交融，界面清晰，可随时分离排出。

emulsion *n.* 乳化物

例：The free water knockout (FWKO) is designed to separate the free water from the oil and <u>emulsion</u>, as well as separate gas from liquid.

游离水脱除器的功能是将游离水从油和乳化物中分离出来，以及进行气液分离。

oil emulsion *np.* 乳化油

例：An <u>oil emulsion</u> is a mixture of oil, water, and an emulsifying agent.

乳化油是油、水和乳化剂的混合物。

emulsifying agent *np.* 乳化剂

例：The properties and amount of <u>emulsifying agent</u> as well as the amount of agitation determine the "stability" of emulsions.

乳状液的"稳定性"取决于其搅动程度及乳化剂的特性和数量。

water–in–oil emulsion *np.* 油包水型乳状液

例：Oil and water can form a <u>water-in-oil emulsion</u>, wherein water is the dispersed phase and oil is the external phase.

油和水可以形成油包水型乳状液，其中水是分散相，油为外相。

oil–in–water emulsion *np.* 水包油乳液

例：On the contrary, they can form an <u>oil-in-water emulsion</u>, where in the oil is the dispersed phase, and water is the dispersion medium.

相反，它们能形成水包油乳液，油是分散相，水是分散介质。

demulsification *n.* 破乳

例：<u>Demulsification</u> is the process of breaking emulsions in order to separate water from oil, which is also one of the first steps in processing the crude oil.

破乳是指破坏乳状液，将水从油中分离出来的过程，也是原油加工的第一步。

demulsifier *n.* 破乳剂

例： Demulsifiers act to neutralize the effect of emulsifying agents.

破乳剂可起到抵消乳化剂的作用。

dispersion *n.* 分散体

例： If no emulsifier is present, the droplets will eventually settle to the bottom causing the smallest interface area. This type of mixture is a true "dispersion".

如果没有乳化剂，水滴最终会沉底，形成最小的界面面积。这类混合液体是真正的"分散体"。

preference for the oil *np.* 亲油

例： An emulsifying agent has a surface active behavior. Some element in the emulsifier has a preference for the oil, and other elements are more attracted to the water.

乳化剂具有表面活性，其某些成分特别亲油，另外一些成分对水更有吸引力。

oil wet *adj.* 亲油的

例： Paraffins and asphaltenes could be dissolved or altered to make their films less viscous so they will flow out of the way on collision or could be made oil wet so they will be dispersed in the oil.

石蜡与沥青质可以溶解或者使其膜黏度降低，这样发生碰撞时就可以向旁边流出，或者具有亲油性，分散在油里。

water wet *adj.* 亲水的

例： Iron sulfides, clays, and drilling muds can be water wet causing them to leave the interface and be diffused into the water droplet.

硫化铁、黏土、钻井液都是亲水的，亲水性使其离开界面并扩散到水滴里。

initial separation *np.* 初次分离

例： If the pressure for initial separation is too high, too many light components will stay in the liquid phase at the separator and be lost to the gas phase at the tank.

若初次分离过程中的压力太高，则分离器中呈液相的轻质组分也较多，同时在该存液罐内，该液相会变为气相而蒸发损失掉。

fractionate *v.* 分馏

例：Often it is more economical to fractionate the liquid into its various components, which have a market value as purity products.

更经济的方法是把液体分馏成不同组分，这些组分单一的产品更有市场价值。

fractionation plant *np.* 分馏装置

例：At the fractionation plant, liquids will be separated into commercial quality products and then delivered to the market by tankers and tank trucks.

在分馏装置中，将液体分离成具有商业价值的产品，然后通过油轮和油罐车运往市场。

physical separation *np.* 物理分离

例：The physical separation of these phases is one of the basic operations in the production, processing, and treatment oil and gas.

在油气开采、加工和处理过程中，基本的作业是对各相流体进行物理分离。

vapor pressure specification *np.* 蒸汽压指标

例：Stabilized liquid will generally have a vapor pressure specification (Reid vapor pressure of < 10 psi), as the product will be injected into a pipeline or transport pressure vessel, which has definite pressure limitations.

稳定处理后的液体要符合蒸汽压指标（雷德蒸汽压小于10psi），因为要注入有一定压力限制的管道或运输压力罐。

produced water disposal *np.* 采出水处理

例：The most common method of dealing with produced water is disposing of it using a piece of equipment such as water pump skids or produced water disposal pumps.

最常用的采出水处理方法是使用水泵支架或采出水处理泵。

primary clarification *np.* 一次沉降

例：Primary clarification is the first step in the water treatment process for removing suspended solids (TSS), oil and grease.

一次沉降是水处理过程中去除悬浮固体（TSS）和油脂的第一步。

water leg *np.* 含水区

例：The oil is skimmed off the surface of the gun barrel and the water exits from the bottom either through a <u>water leg</u> or an interface controller and dump valve.

原油从油水分离器（沉降罐）表面撇去，水通过含水区或者油水界面控制器及排水阀从底部排出。

the maximum hydrocarbon content *np.* 最大含烃量

例：In offshore areas where discharge to the sea is allowed, the governing regulatory body specifies <u>the maximum hydrocarbon content</u> in the water that may be discharged overboard.

监管机构对近海排放区船上可排放到海里的含油污水规定了最大含烃量。

desalt *n.* 脱盐

例：Crude oil enters an atmospheric distillation unit and starts at <u>desalting</u>.

原油进入常压蒸馏装置，并开始脱盐。

primary treating *np.* 初级处理

例：Produced water will always have some form of <u>primary treating</u> prior to disposal.

采出水排放前总要进行某种形式的初级处理。

secondary treating *np.* 二次处理

例：Depending upon the severity of the treating problem, <u>secondary treating</u> utilizing a CPI, crossflow separator, or a flotation unit may be required.

根据具体处理难度，可能需要使用波纹板隔油池、交叉流分离器或浮选设备进行二次处理。

volatility *n.* 挥发度

例：These steps remove enough light hydrocarbons to produce a stable crude oil with the <u>volatility</u> to meet sales criteria.

这些工序去除了轻烃，足以生产出性能稳定且具有挥发性的原油，符合销售标准。

dew point *np.* 露点

例：In some locations it may be necessary to remove the heavier hydrocarbons to lower the hydrocarbon <u>dew point</u>.

某些油气产区为了降低烃的露点，可能需要去掉些重烃成分。

sweeten *v.* 脱硫

例：Contaminants such as H_2S and CO_2 may be present at levels higher than those acceptable to the gas purchaser. If this is the case, then additional equipment will be necessary to "<u>sweeten</u>" the gas.

天然气中的污染物，诸如 H_2S 及 CO_2 的含量，可能高于天然气销售所允许的范围。若遇到这种情况，就需要用其他处理设备给天然气脱硫。

basic sediment and water (BS and W) *np.* 底部沉淀物及水分

例：Most oil contracts specify a maximum percent of <u>basic sediment and water</u> that can be present in the crude.

多数石油合同都规定了原油中可残留的底部沉淀物及水分的最高含量。

sampler *n.* 取样器

例：The <u>sampler</u> receives a signal from the meter to assure that the sample size is always proportional to flow even if the flow varies.

取样器从流量计接收信号，以确保所取油样与油流成比例，即使油流发生变化。

positive displacement meter *np.* 正排流量计

例：The liquid then flows through a <u>positive displacement meter</u>.

然后该油流流经正排流量计。

interstage pressure *np.* 级间压力 / 门限压力

例：Depending on the pressure at the plant gate, the next step in processing will either be inlet compression to an "<u>interstage pressure</u>", or be dew point control and natural gas liquid recovery.

根据处理厂进口压力的大小，下一步或者是将气体增压到"门限压力"，或是控制露点，回收天然气凝析液。

hydrocarbon dew point *np.* 烃露点

例：Hydrocarbon dew point or hydrocarbon liquid recovery involves cooling the gas and condensing out the liquids.

烃露点控制或液态烃回收涉及气体冷却和液体凝结析出两个过程。

inhibition/cooling/condensation process *np.* 抑制／冷却／凝析作用

例：Hydrocarbon dew point control can be either dehydration followed by cooling, condensation or by a combination of inhibition, cooling and condensation processes.

烃露点控制既可以通过脱水然后冷却、凝析完成，也可以通过抑制、冷却以及凝析三作用者共同完成。

two-stage separation *np.* 两级分离

例：Two-stage separation is usually used for low GOR and low well stream pressure, three-stage separation is used for medium to high GOR and intermediate inlet pressure, and four-stage separation is used for high GOR and a high pressure well stream.

两级分离通常用于低气油比和低井流压力；三级分离用于中至高气油比和中井流压力；四级分离用于高气油比和高井流压力。

multistage separation *np.* 多级分离

例：Multistage separation is applied to achieve good separation between gas and liquid phases and maximizing hydrocarbon liquid recovery.

应用多级分离可使气液两相更好地分离，以便烃类液体回收率最大化。

stage separation process *np.* 多级分离工艺

例：In a stage separation process the light hydrocarbon molecules that flash are removed at relatively high pressure, keeping the partial pressure of the intermediate hydrocarbons lower at each stage.

在多级分离过程中，闪蒸后的轻烃分子在高压下会分离出来，使每级分离过程中的中间烃类分压处于较低状态。

production separator *np.* 生产分离器

例：Production separator is used in crude oil treatment and production for efficient separation of oil, gas, and water.

生产分离器用于原油处理和生产中,可有效分离石油、天然气和水。

two-phase separator *np.* 两相分离器

例:<u>Separators</u> are classified as "<u>two-phase</u>" if they separate gas from the total liquid stream.

若所用分离器只是将天然气从总的液流中分离出来,这种分离器就叫两相分离器。

three-phase separator *np.* 三相分离器

例:A <u>three-phase separator</u> can separate the liquid stream into its crude oil, gas and water components.

三相分离器可以将液相中的原油、天然气和水等组分分离开来。

horizontal separator *np.* 卧式分离器

例:Normally, <u>horizontal separators</u> are operated half full of liquid to maximize the surface area of the gas liquid interface.

通常,卧式分离器工作时液体占一半,以保证气—液界面的表面积最大。

vertical separator *np.* 立式分离器

例:In the <u>vertical separator</u>, the gas flows over the inlet diverter and then vertically upward toward the gas outlet.

在立式分离器中,天然气经过入口分流器,然后垂直向上流向气体出口。

spherical separator *np.* 球型分离器

例:<u>Spherical separators</u> are a special case of a vertical separator where there is no cylindrical shell between the two heads.

球型分离器可看作没有中间圆筒壳而只有上、下两个球型帽的立式分离器。

free-water knockout *np.* 游离水脱离器

例:After the gas has passed through a <u>free-water knockout</u>, glycol is sprayed in small droplets into the gas to absorb the water.

天然气经过游离水脱离器后,乙二醇以小滴形式喷入天然气中吸收水分。

API separator *np.* API 油水分离器

例：The <u>API separator</u> is normally the first, and most important, wastewater treatment step in petroleum refineries.

API 油水分离器通常是处理炼油废水的第一步，也是最重要的一步。

initial separator *np.* 初级分离器

例：<u>Initial separator</u> can be used for oil-water primary separation, mainly because the density of oil is less than water and oil floats on water, the separation equipment moves back and forth.

初级分离器可用于油水初级分离，主要由于油密度小于水密度，油浮于水上，分离设备来回移动。

oil and gas separator *np.* 油气分离器

例：An <u>oil and gas separator</u> is a vertical or horizontal vessel that producers use to separate the elements of oil, gas, water in the fluid stream.

油气分离器是生产者用来分离流体中油、气、水组分的垂直或水平容器。

hydrocyclone *n.* 旋流分离器

例：Facilities may include sedimentation basins or tanks, <u>hydrocyclones</u>, filters.

这些处理装置可能包括沉降池或沉降罐、旋流分离器以及过滤器。

gravity separator *np.* 重力分离器

例：<u>Gravity separators</u> are pressure vessels that separate a mixed-phase stream into gas and liquid phases that are relatively free of each other.

重力分离器是压力容器，可将混合相分离为相对游离的气相和液相。

high-pressure separator *np.* 高压分离器

例：To recover the gas fractions produced in the separators operating at medium pressure and low pressure, it is necessary to recompress them at the pressure of the <u>high-pressure separator</u>.

中低压下运行的分离器所产生的气体组分，需要在高压分离器压力下再压缩才能回收。

centrifugal or cyclone separator *np.* 离心或旋风分离器

例：In centrifugal or cyclone separators, centrifugal forces act on droplet at forces several times greater than gravity as it enters a cylindrical separator.

在离心或旋风分离器中，当液滴进入圆柱形分离器时，离心力对液滴的作用力是重力的几倍。

twister supersonic separator *np.* 扭转式超声分离器

例：The twister supersonic separator is a unique combination of known physical processes, combining expansion, cyclonic gas/liquid separation, and recompression process steps in a compact, tubular device to condense and separate water and heavy hydrocarbons from natural gas.

扭转式超声分离器独特地综合了一系列物理过程，包括膨胀、旋流式气—液分离，以及再压缩等处理步骤，采用紧凑的管状装置，使天然气凝析并分离出水和重烃组分。

gravity settling *np.* 重力沉降

例：Three principles used to achieve physical separation of gas and liquids or solids are momentum, gravity settling, and coalescing.

实现气体、液体或固体物理分离的三个原理是动量学、重力沉降和聚结。

inlet diverter *np.* 进口分流器

例：In most designs, the inlet diverter contains a downcomer that directs the liquid flow below the oil/water interface.

在大部分设计中，进口分流器带有一个降液管，引导液体在油水界面下流动。

downcomer *n.* 降液管

例：A downcomer is required to transmit the liquid collected through the oil-gas interface so as not to disturb the oil-skimming action taking place.

降液管输送从油气界面收集到的液体，以避免干扰撇油作业。

spreader *n.* 铺散器

例：The spreader or downcomer outlet is located at the oil–water interface.

铺散器或降液管出口位于油水界面处。

particulate loading *np.* 携带颗粒

例：Centrifugal separators are also extremely useful for gas streams with high particulate loading.

离心分离器分离携带颗粒较多的气流也是相当有效的。

demisting device/demister/mist extractor *np.* 除雾设备

例：The liquid level can affect the pressure drop for the downcomer pipe (from the demister), therefore affecting demisting device drainage.

液位能影响降液管压降，因而影响除雾设备的排驱。

scrubber *np.* 净气器

例：Separators are sometimes called "scrubbers" when the ratio of gas rate to liquid rate is very high.

当气液流速比非常高时，分离器有时也称为净气器。

gas/oil ratio (GOR) *np.* 气油比

例：In practice, the number of stages normally ranges between two and four, which depends on the gas/oil ratio (GOR) and the well stream pressure.

实际上，分离级数通常介于二到四级，取决于气油比和井流压力。

twister system *np.* 气旋式系统

例：The compact and low weight twister system design enables debottlenecking of existing space and weight-constrained platforms.

低重量气旋式系统能解决目前平台空间和重量受限的瓶颈问题。

slug catcher *np.* 段塞流捕集器

例：The slug catcher is designed to separate gas, hydrocarbon condensate, and inlet water.

段塞流捕集器是用来分离气体、凝析烃和井口水的。

vessel type slug catcher *np.* 容器式段塞流捕集器

例：Pipe-type slug catchers are frequently less expensive than vessel type slug catchers of the same capacity due to thinner wall requirements of smaller pipe diameter.

相同容量的管式段塞流捕集器往往比容器式更便宜，因为管壁更薄，管径更小。

multiple pipe-type slug catcher *np.* 多管型段塞流捕集器

例：The manifold nature of multiple pipe-type slug catchers also makes possible the later addition of additional capacity by laying more parallel pipes.

多管型段塞流捕集器可重叠布置，这样可以铺设更多平行管道，从而增加容量。

condensate slug *np.* 凝析油段塞

例：When doing condensate/water separation at the slug catcher itself, we have to allow separately for the maximum condensate slug and the maximum water slug in order to ensure continuous level control.

用段塞流捕集器进行凝析油/水分离时，既要保证凝析油段塞最大化，也要保证水段塞最大化，以确保液位控制的连续性。

finger *n.* 指型管

例：Separation of gas and liquid phases is achieved in the first section of the fingers.

气体和液体两相的分离，是在指型管的第一段实现的。

gas riser *np.* 天然气立管

例：The length of the intermediate section is minimal such that there is no liquid level beneath the gas riser when the slug catcher is full, i.e., storage section completely full.

中间段的长度是最短的，保证液体段塞捕集器充满时，天然气立管下没有液面，即储存区完全充满。

slug volume *np.* 段塞储集体积

例：The length of the storage section ensures that the maximum slug volume can be retained without liquid carryover in the gas outlet.

储存段的长度要确保最大段塞储集体积，不能让液体进入天然气出口。

main liquid collection header *np.* 主液体收集汇管箱

例：During normal operations, the normal liquid level is kept at around the top of the riser from each finger into the main liquid collection header, which is equivalent to approximately a 5-min operation of the condensate stabilization units at maximum capacity.

正常运行期间，正常液面保持在立管的顶部上下，液体从每个指型管流入主液体收集汇管箱，这相当于凝析稳定装置以最大容量运行大约 5 分钟。

flash v. 闪蒸

例：The liquid is first <u>flashed</u> at an initial pressure and then flashed at successively lower pressures two times before entering the stock tank.

液体首先在初始压力下闪蒸，接着保持在低压下闪蒸两次，再进入存液罐。

vacuum evaporation np. 真空蒸发

例：The <u>vacuum evaporation</u> is the only way to separate oil from water without having to pre-treat the effluent or requiring subsequent processes, as the water produced is of higher quality and can be reused directly.

真空蒸发是将油与水分离的唯一方法，因为产出水质量更高，可以直接重复使用，无须对排出水进行预处理或后续处理。

vacuum dehydrators np. 真空脱水器

例：<u>Vacuum dehydrators</u> are oil and gas filtration systems used to remove water contaminants from oil or gas inputs.

真空脱水器是一种油气过滤系统，用于去除油气输入中的水杂质。

molecular sieve dryer np. 分子筛网烘干机

例：The water is removed by passing the feed stream through a silica gel or <u>molecular sieve dryer</u>.

原料经过进料管时，硅胶或分子筛网烘干机将水脱除。

suspended solid np. 悬浮固体

例：<u>Suspended solids</u> refer to small solid particles which remain in suspension in water as a colloid.

悬浮固体指固体小颗粒，以胶体形式悬浮在水中。

dissolved impurity np. 溶解杂质

例：<u>Dissolved impurities</u> are in a liquid having only one phase, so such impurities can be removed only by phase change such as precipitation, adsorption, distillation.

液体只有一个相态，去除其中的溶解杂质只能通过相变法，如沉淀、吸附、蒸馏等。

sludge *n.* 油泥

例：<u>Sludge</u> is a thick, viscous emulsion containing oil, water, sediment and residue that forms because of the incompatibility of certain native crude oils and strong inorganic acids used in well treatments.

油泥是一种黏稠的乳液，含有油、水、沉淀物和残渣，由于某些天然原油与井处理中使用的强无机酸不相容而形成。

air flotation *np.* 气浮除油

例：<u>Air flotation</u> is an efficient way to extract oil from wastewater.

气浮除油是从废水中提取石油的有效方法。

dissolved air flotation (DAF) *np.* 溶气气浮法

例：The <u>dissolved air flotation (DAF)</u> method has an important role in the removal of hydrocarbons, as well as in the protection of the biological treatment.

溶气气浮法（DAF）在去除烃类和保护生物处理方面发挥了重要作用。

hydrocyclone *n.* 水力旋流器

例：<u>Hydrocyclone</u> is suitable for high concentrations of oils.

水力旋流器适用于处理高浓度油品。

plate coalescer *np.* 板式聚结器

例：The working principle of all <u>plate coalescers</u> depends on gravity separation to allow the oil droplets to rise to a plate surface where coalescence and capture occur.

板式聚结器的工作原理就是利用重力分离，使废水中的油滴浮升到聚结器的板表面，并且发生聚结和油滴捕集。

gas scrubber *np.* 气体洗涤器

例：Separators are sometimes called "<u>gas scrubbers</u>" when the ratio of gas rate to liquid rate is very high.

当气油产出比很高时，分离器称为气体洗涤器。

trap *n.* 捕集器

例：Some operators use the term "<u>traps</u>" to designate separators that handle flow directly from wells.

某些操作人员称直接处理出油流的分离器为"捕集器"。

liquid dump valve *np.* 液体放空阀

例：The liquid dump valve is regulated by a level controller.

该液体放空阀可由液面控制器调节。

level controller *np.* 液面控制器

例：The level controller senses changes in liquid level and controls the dump valve accordingly.

液面控制器感知液面变化，相应地控制放空阀。

gas blanket *np.* 气体覆盖层

例：A gas blanket is provided to assure that there is always enough pressure in the treater so the water will flow to water treating.

原油处理设备中有气体覆盖层，以保证设备中有足够的压力，使脱出水流入污水处理设备中。

mist extractor/mist eliminator *np.* 除雾器

例：Gas goes through the mist extractor section before it leaves the vessel.

天然气流经除雾器除水后，离开分离器。

cyclone separator *np.* 旋转分离器

例：Cyclone separators are designed to operate by centrifugal force. These designs are best suited for fairly clean gas streams.

旋转分离器是利用离心力运转，能产出非常洁净的气流。

centrifugal force *np.* 离心力

例：The swirling action of the gas stream as it enters the scrubber separates the droplets and dust from the gas stream by centrifugal force.

气流进入分离器，由于离心力旋转作用，气雾和灰尘分离。

dehydration tower *np.* 脱水塔

例：These units are commonly used to recover glycol carryover downstream of a dehydration tower.

这些装置通常用于回收脱水塔下游的乙二醇。

two-barrel separator *np.* 双筒式分离器

例：Two-barrel separators are common where there is a very low liquid flow rate. In these designs the gas and liquid chambers are separated.

处理小流量流体时，通常采用双筒式分离器，其气体筒和液体筒是分开的。

filter separator *np.* 过滤分离器

例：Another type of separator that is frequently used in some high-gas/low-liquid flow applications is a filter separator. These can be either horizontal or vertical in configuration.

另有一种适用于高气流、低液流的分离器，叫过滤分离器，有卧式和立式两种结构。

flocculate *v.* 絮凝

例：In the context of heavy oil, asphaltenes are known to flocculate at the molecular level (before precipitation) and in the precipitated state.

在稠油中，沥青质在沉淀前的分子水平和沉淀状态下会发生絮凝。

buffer tank *n.* 稳压罐

例：Buffer tank is a targeted, horizontal, cylindrical tank that changes the direction of fluid flow downstream of the choke and serves to direct flow to the flare line or gas buster.

稳压罐是定向水平圆柱形罐，可以改变节流器下游的流体流向，引导流体流向火炬管或气体分离器。

pressure/vacuum valve *np.* 压力/真空阀

例：All tanks should have a pressure/vacuum valve with flame arrester and gas blanket to keep a positive pressure on the system and exclude oxygen.

为保持系统正压，排除氧气，所有储罐必须装有带阻火器的压力/真空阀及气体覆盖层。

pressure controller *np.* 压力控制器

例：The pressure controller senses changes in the pressure in the separator and sends a signal to either open or close the pressure control valve accordingly.

压力控制器感知分离器中的压力变化，并发出相应信号控制压力阀开关。

flame arrester *np.* 阻火器

例：A <u>flame arrester</u> is a device fitted to the opening of an enclosure to allow gases, liquids, etc. to pass through.

阻火器安装在封罩口处，可使气体、液体等通过。

hydrometer *n.* 石油密度计

例：A <u>hydrometer</u> is an instrument used for measuring density or relative density of liquids based on the concept of buoyancy.

石油密度计是基于浮力原理来测量液体密度或相对密度的仪器。

water finding paste *np.* 试水膏

例：<u>Water finding paste</u> is normally used in the petroleum industry to monitor and detect moisture levels in various fuel types.

试水膏通常用于石油工业，用于监测和检测各种燃料的湿度。

gauging paste *np.* 量油膏

例：<u>Gauging paste</u> indicates levels of gasoline, naphtha, kerosene, crude and refined oils, jet fuels and more in fuel storage tanks.

量油膏可检测燃料储罐中的汽油、石脑油、煤油、原油和精炼油、航空燃料等的含量。

vibratory shear enhanced processing (VSEP) *np.* 振动膜过滤系统

例：The use of <u>vibratory shear enhanced processing</u> could potentially allow for the production of high-quality water from any oil emulsion in water.

使用振动膜过滤系统，可以从水包油乳状液中产出高品质的水。

filter media *np.* 滤介质

例：A <u>filter media</u> is one of the most essential parts of any oil and gas filtration system.

滤介质是油气过滤系统中最重要的部件之一。

simplex filter *np.* 单纯形过滤器

例：<u>Simplex filters</u> are the most basic filters with a compact design and a simple filtration process.

单纯形过滤器是最基本的过滤器，设计紧凑且过滤过程简单。

duplex filter *np.* 双联过滤器

例：The duplex filters are one of the commonly used types in many oil and gas filtration systems.

许多油气过滤系统常用双联过滤器。

coalescing filter *np.* 凝聚过滤器

例：By filtering with particulate and coalescing filters, solids and liquids, such as sand, dust, and water, are removed.

通过微粒过滤器和凝聚过滤器对天然气进行过滤，可以去除天然气中的固体和液体，如沙子、灰尘和水。

multi bag filter *np.* 多袋式过滤器

例：Multi bag filters are used in a variety of applications across multiple industries including industrial manufacturing, oil and gas, municipal water and other power and utilities applications.

多袋式过滤器用于多种行业，如工业制造、石油和天然气、市政供水以及其他电力和公用事业。

water spray cooling system *np.* 冷却水喷淋系统

例：Water spray cooling systems can reduce damages to the tank on fire and reduce the risk of escalation and delay involvement of adjacent tanks.

油罐着火时，油罐冷却水喷淋系统可以减少火灾损失，防止火情恶化或波及邻近油罐。

flotation unit *np.* 浮选设备

例：Under other circumstances, a full system of plate coalescers, flotation units, and skim piles may be needed.

其他情况下，可能需要一套完整的处理系统，包括板式聚结器、浮选设备和隔油管桩等。

gravity separation *np.* 重力分离

例：All of these devices employ gravity separation techniques.

所有这些设备都采用重力分离技术。

vertical skimmer *np.* 立式隔油罐

例：In vertical skimmers the oil droplets must rise upward countercurrent to the downward flow of the water.

在立式隔油罐中，水向下流，而油滴必须逆流而上。

inlet spreader *np.* 入口分流器

例：Some vertical skimmers have inlet spreaders and outlet collectors to help even the distribution of the flow.

有些立式隔油罐还装有入口分流器和出口集水器，使水流均匀分布。

quiet zone *np.* 滞留区

例：In the quiet zone between the spreader and the water collector, some coalescence can occur and the buoyancy of the oil droplets causes them to rise counter to the water flow.

分离器与集水器之间的滞留区可能会发生聚结，油滴浮力使其与水流逆向浮升。

horizontal skimmer *np.* 卧式隔油罐

例：In horizontal skimmers, the oil droplets rise perpendicular to the flow of the water.

在卧式隔油罐中，油滴浮升方向与水流方向垂直。

sand drain *np.* 排砂

例：Sand and other solid particles must be handled in vertical vessels with either the water outlet or a sand drain at the bottom.

处理砂子和其他固体颗粒时，须将出水口或排砂口安排在立式容器底部。

atmospheric vessel *np.* 常压容器

例：The water must be dumped to a higher level for further treating and a pump would be needed if an atmospheric vessel were installed.

水必须排到更高标高才能进一步处理，如果安装常压容器可能需要用泵。

gas venting *np.* 气体放空

例：Due to the potential danger from overpressure and potential gas venting problems associated with atmospheric vessels, pressure vessels are preferred.

由于常压容器存在潜在的超压危险和气体放空问题，应当优先选用压力容器。

short circuiting *np.* 短路

例：Skimmers with long residence times require baffles to attempt to distribute the flow and eliminate short circuiting.

隔油罐内停留时间较长时，需要用折流板改善流体分布状况，消除短路。

disposal pile *np.* 排放管桩

例：Offshore, produced water can be piped directly overboard after treating, or it can be routed through a disposal pile or a skim pile.

海上含油污水经处理后可直接排出采油船外，或排入排放管桩及隔油管桩。

streamline *n.* 层流

例：Below a certain critical velocity, the flow is streamline.

当流速低于一定的临界速度时，液流便是层流。

loss of head *np.* 水头损失

例：An accurate estimation of the probable loss of head in a pipe is important.

准确估计管道中可能产生的水头损失非常重要。

hot oil flushing *vp.* 热洗 / 热油冲洗

例：Hot oil flushing is done to clear any contaminants present in the pipe and oil after installation.

热油冲洗是为了清除管道安装后管道和油中的污染物。

oil-injected screw compressor *np.* 注油螺杆压缩机

例：An oil-injected screw compressor has an oil separating mechanism integrated with a compressor and hence is made compact in size.

注油螺杆压缩机具有分离机制，与压缩机集成在一体需要设计尺寸小巧。

grease trap *np.* 隔油池

例：The definition of a grease trap is a trap in a drain or waste pipe to prevent grease from passing into a sanitary sewer lines and system.

隔油池指的是排水渠或废水管内的隔油池，以防止油脂流入卫生污水渠管道及系统。

asphalt coating *np.* 石油沥青防腐层

例：Asphalt coating is one of the most important anticorrosive coatings.

石油沥青防腐层是重要的防腐涂料之一。

high-voltage fuse *np.* 高压熔断器

例：High-voltage fuses are made to prevent this conductive film from forming when the fuse blows, a common device in oil and gas process.

高压熔断器用于防止熔断时形成这种导电膜，是油气处理中的常用装置。

safety instrumented system *np.* 安全仪表系统

例：The safety instrumented system is an important part of automatic control in oil plants.

安全仪表系统是石油工厂自动控制中的重要组成部分。

heat exchanger *np.* 热交换设备

例：Unregulated industrial processes can result in detrimental temperature spikes affecting machinery, transport channels, and even the product itself, so in order to maintain optimal thermal conditions in oil and gas systems, heat exchangers are often employed.

工业流程不规范会导致温度峰值过高，影响机械运作、运输管道以及石油产品，所以在石油和天然气工业中，常使用热交换设备来保持最佳的热力条件。

plate-type heat exchangers *n.* 板式换热器

例：Plate-type heat exchangers can help increase heat process efficiency and lower costs for onshore and offshore oil and gas production.

板式换热器有助于提高热加工效率，并降低陆上和海上石油和天然气生产成本。

shell and tube heat exchanger *np.* 壳管式热交换器

例：A shell and tube heat exchanger is a mechanical device that uses a tube mounted in a cylindrical housing to exchange heat between two hydraulic fluids through thermal contact.

壳管式热交换器是一种机械装置，借助置于圆柱形外壳的管子，通过热接触在两种液压油间交换热量。

reconcentrator n. 重沸器

例：<u>Reconcentrator</u> is widely used in petroleum, chemical, metallurgy, and other industries of heat transfer equipment, plays a particularly important role in the petrochemical field.

重沸器是广泛应用于石油、化工、冶金等行业的换热设备，在石油化工领域中发挥着极其重要的作用。

Section IV

Natural Gas Processing
天然气处理

natural gas processing plant *np.* 天然气处理厂

例：Natural gas processing plants clean raw natural gas by separating impurities and various nonmethane hydrocarbons and fluids to produce what is known as pipeline quality dry natural gas.

通过分离杂质及各种非甲烷碳氢化合物和流体来清洁原料天然气，天然气处理厂生产出符合管道标准的干气。

liquefied natural gas plant *np.* 液化天然气厂

例：The costs of building a liquefied natural gas plant have lowered since the mid-1980s for the improvement of thermodynamic efficiencies.

自20世纪80年代中期以来，由于热动力学的发展，液化天然气厂建设成本下降。

produced gas stream *np.* 采出气

例：These processes can be used directly on the produced gas stream.

这些工艺可直接用在采出气上。

unprocessed gas stream *np.* 未经处理的天然气

例：Hydrate inhibition using chemical inhibitors is still the most widely used method for unprocessed gas streams.

采用化学抑制剂抑制水合物仍然是运输未经处理的天然气使用的最为广泛的方法。

wet gas *np.* 湿气

例：Wet gas refers to the raw, unprocessed natural gas that contains large amounts of associated natural gas liquids.

湿气是指未加工的天然气，含有大量伴生的天然气凝析液。

overhead gas *np.* 塔顶气

例：Overhead gas from the three-phase separator is recompressed where necessary for use as fuel gas.

三相分离器中分离出的塔顶气，用作燃料气时需要再次压缩。

acid gas *np.* 酸性气体

例：H_2S when combined with water forms a weak sulfuric acid, whereas CO_2 and water form carbonic acid, thus the term "acid gas".

硫化氢和水混合形成弱硫酸，而二氧化碳和水混合后形成碳酸，因此称为酸性气体。

sour gas *np.* 酸气

例：Strictly speaking, a sour gas is any gas that specifically contains hydrogen sulfide in significant amounts.

严格来说，酸气是指含有大量硫化氢的气体。

sweet gas *np.* 甜气

例：Natural gas with H_2S or other sulfur compounds present is called "sour gas", whereas gas with only CO_2 is called "sweet gas".

含硫化氢和其他含硫化合物的天然气称为"酸气"，而只含二氧化碳的气体称为"甜气"。

retrograde condensate gas *np.* 反凝析气

例：The second situation to motivate maximum condensate recovery occurs when processing retrograde condensate gas.

第二种希望凝析油采收率最大化的情况是在处理反凝析气时。

dry gas *np.* 干气

例：Natural gas is considered "dry gas" when it is almost pure methane, having had most of the other commonly associated hydrocarbons removed.

天然气脱除大部分其他常见的伴生烃，组分几乎接近纯甲烷，此时的天然气称为"干气"。

rich gas *np.* 富气

例：The terms rich gas and lean gas, as used in the gas-processing industry, are not precise indicators of gas quality but only indicate the relative amount of natural gas liquids in the gas stream.

在天然气处理中，富气和贫气的概念并不能准确说明天然气质量的好坏，而是反映天然气中液态烃相对含量的多少。

lean gas *np.* 贫气

例：Depending on its content of heavy components, natural gas can be considered as rich (5 or 6 gallons or more of recoverable hydrocarbons per cubic feet) or lean gas (less than 1 gallon of recoverable hydrocarbons per cubic feet).

根据重组分的含量，天然气可以分为富气（不小于 5~6gal/ft^3 可采油气量）和贫气（小于 1gal/ft^3 可采油气量）。

natural gas liquid（NGL） *np.* 天然气凝析液

例：The separated gaseous stream is traditionally very rich in natural gas liquids (NGLs).

分离后的气流通常富含天然气凝析液。

liquefied petroleum gas *np.* 液化石油气

例：The gas produced in the separators and the gas that comes out of solution with the produced crude oil are then treated to separate out the NGLs that are treated in a gas plant to provide propane and butane or a mixture of the two (liquefied petroleum gas).

分离器中分离出的气体和原油生产中来自溶解气的气体，经过处理后分离出天然气凝析液，凝析液再经处理分离出丙烷和丁烷，或者二者的混合物，即液化石油气。

compressed natural gas (CNG) *np.* 压缩天然气

例：Compressed natural gas is used in some countries for vehicular transport as an alternative to conventional fuels (gasoline or diesel).

一些国家将压缩天然气（CNG）用于交通，以替代常规燃料（如汽油或柴油）。

strainer /gas eliminator *np.* 天然气过滤器

例：The crude first flows through a strainer /gas eliminator to protect the meter and to assure that there is no gas in the liquid.

原油首先要流经天然气过滤器，以保护流量计，并保证油流中不含气。

gas dehydrator *np.* 天然气脱水器

例：Gas dehydrator is widely used in natural gas treatment plant as a common process because water and hydrocarbons can form hydrates, which may block valves and pipelines.

由于水和碳氢化合物会形成水合物，堵塞阀门和管道。因此，天然气脱水器作为常用设备在天然气处理厂被广泛应用。

hydrate *n.* 水合物

例：Removing most of the water vapor from the gas is required by most gas sales contracts, because it prevents hydrates from forming when the gas is cooled in the transmission and distribution systems and prevents water vapor from condensing and creating a corrosion problem.

大多数天然气销售合同都规定天然气须脱除大部分水气，以防止天然气在输配系统中冷却，水气冷凝形成水合物，造成腐蚀问题。

wet glycol *np.* 含水醇类

例：The wet glycol leaves from the base of the tower and flows to the reconcentrator (reboiler) by way of heat exchangers, a gas separator, and filters.

含水醇类从塔的下部流出后，通过换热器、气体分离器和过滤器流入重沸器。

dry glycol *np.* 脱水醇类

例：The dry glycol is then pumped back to the contact tower.

脱水醇类接着泵入接触塔。

glycol dehydration unit *np.* 乙二醇脱水装置

例：In order to reduce the size of the solid desiccant dehydrator, a glycol dehydration unit is often used for bulk water removal.

为了减小固体干燥器的体积，经常使用乙二醇脱水装置去除大量水分。

natural gas cooler *np.* 天然气冷却装置

例：Natural gas coolers are able to eliminate the negative operating effects of high gas and liquid temperature.

天然气冷却装置能够消除气液高温作业的负面影响。

two-stage filtration *np.* 两级过滤

例：The natural gas goes through a two-stage filtration through a compressor station.

天然气在压缩站经过两级过滤。

gas desulfurization *np.* 气体脱硫

例：In the process of flue gas desulfurization, the dissolution of limestone is an important reaction control step.

气体脱硫过程中，石灰石溶解是一个重要的反应控制步骤。

sulfur recovery *np.* 硫回收

例：Sulfur recovery units face catalyst deactivation due to aromatic contaminants in acid gas.

由于酸性气体含有芳香族污染物，硫回收装置面临催化剂失活的问题。

absorbent *n.* 吸收剂

例：Absorption differs from adsorption in that it is not a physical–chemical surface phenomenon, but an approach in which the absorbed gas is ultimately distributed throughout the absorbent (liquid).

吸收不同于吸附，因其不是物理化学表面现象，而是一种使吸收气最终完全分散于吸收剂（液体）中的方法。

absorbing media *np.* 吸收介质

例：Common absorbing media used are water, aqueous amine solutions, sodium carbonate, and nonvolatile hydrocarbon oils, depending on the type of gas to be absorbed.

常用的吸收介质有水、醇胺水溶液、氢氧化钠、碳酸钠溶液和非挥发性的液态烃，根据所处理气体类型选择相应的吸收剂。

absorption *n.* 吸收作用

例：<u>Absorption</u> is achieved by dissolution (a physical phenomenon) or by reaction (a chemical phenomenon).

吸收作用通过溶解作用（物理现象）或化学反应（化学现象）实现。

absorber *n.* 吸收器

例：Then the absorbent is stripped of the gas components (regeneration) and recycled to the <u>absorber</u>.

然后吸收剂解析气体成分并再循环至吸收器中。

adsorption *n.* 吸附法

例：Among all, <u>adsorption</u> is one of the methods that is focused by researchers as it is proven to give a promising result, highly efficient and economically viable for marine oil pollution issues.

其中，吸附法是目前研究人员关注的一种方法。研究证明，利用该方法处理海洋石油污染，效果好、效率高、经济上可行。

activated carbon adsorption *np.* 活性炭吸附法

例：<u>Activated carbon adsorption</u> can provide an extremely important economic contribution to the operations of oil and gas refineries.

使用活性炭吸附法可为炼油厂带来巨大经济效益。

multiple absorption column *np.* 多级吸收器

例：The process design will vary and, in practice, may employ <u>multiple absorption columns</u> and multiple regeneration columns.

工艺设计将多元化，实践中可能使用多级吸收器和多级再生器。

liquid absorption process *np.* 液体吸收方法

例：<u>Liquid absorption processes</u> are classified either as physical solvent processes or as chemical solvent processes.

液体吸收方法分为物理溶剂法和化学溶剂法。

gas–liquid contactor *np.* 气液发生器

例：Usually, the <u>gas–liquid contactor</u> designs that are employed are plate columns or packed beds.

常用的气液发生器是平顶筒或者填充反应床。

regeneration (desorption) *n.* 再生（解析）

例：Regeneration (desorption) can be brought about by the use of reduced pressures and/or high temperatures, whereby the acid gases are stripped from the solvent.

通过降压或升温进行再生（解析），使酸气从溶剂中分离出来。

amine washing of natural gas *np.* 醇胺法除酸气

例：Amine washing of natural gas involves chemical reaction of the amine with any acid gases with the liberation of an appreciable amount of heat and it is necessary to compensate for the absorption of heat.

醇胺法除酸气过程中，醇胺和酸性气体发生化学反应释放出大量热量，用以补偿吸收的热量。

mechanical refrigeration *np.* 机械制冷

例：Mechanical refrigeration is the simplest and most direct process for NGLs recovery.

机械制冷法是最简单和最直接的天然气凝析液回收方法。

lean oil absorption *np.* 贫油吸收法

例：Lean oil absorption is the oldest and least efficient process to recover NGLs.

贫油吸收法是最古老的、效率最低的天然气凝析液回收方法。

chemical adsorption process *np.* 化学吸附过程

例：Chemical adsorption processes adsorb sulfur dioxide onto a carbon surface where it is oxidized and absorbs moisture to give sulfuric acid impregnated into and on the adsorbent.

化学吸附过程是将二氧化硫吸附到炭表面，在炭表面发生氧化并吸收水分，生成浸入吸附剂中的硫酸。

chemical absorption process *np.* 化学吸收法

例：Chemical absorption processes with aqueous alkanolamine solutions are used for treating gas streams containing hydrogen sulfide and carbon dioxide.

化学吸收法利用链烷胺水溶液，处理含有二氧化碳和硫化氢的天然气流。

carbonate washing *np.* 碳酸洗

例：<u>Carbonate washing</u> is a mild alkali process for emission control by the removal of acid gases (such as carbon dioxide and hydrogen sulfide) from gas streams.

碳酸洗是为了控制排放，清除气流中酸气（如二氧化碳和硫化氢）的弱碱处理过程。

water washing *np.* 水洗

例：The inlet diverter assures that little gas is carried with the liquid, and the <u>water washing</u> assures that the liquid does not fall on top of the gas/oil or oil/water interface.

进口分流器可确保液体中基本不含气体，水洗可确保油气或油水界面上没有液体滴落。

organic physical solvent *np.* 有机物理溶剂

例：Methanol was the first commercial <u>organic physical solvent</u> and has been used for hydrate inhibition, dehydration, gas sweetening, and liquids recovery.

甲醇是第一个商用的有机物理溶剂，并已用于水合物抑制、脱水、天然气脱硫和液体回收。

desorption step *np.* 解吸作用

例：Water washing, in terms of the outcome, is analogous to washing with potassium carbonate, and it is also possible to carry out the <u>desorption step</u> by pressure reduction.

从结果来看，水洗与用碳酸钾洗涤效果相同，在压力逐渐下降的情况下，也能发生解吸作用。

acid gas removal process *np.* 脱酸气方法

例：As currently practiced, <u>acid gas removal processes</u> involve the chemical reaction of the acid gases with a solid oxide or selective absorption of the contaminants into a liquid that is passed countercurrent to the gas.

目前使用的脱酸气方法包括：酸气和固体氧化物的化学反应，或有选择性地将杂质吸收到与气体逆向流动的液体中。

membrane separation process *np.* 薄膜分离法

例：The membrane separation process offers a simple and low-cost solution for removal and recovery of heavy hydrocarbons from natural gas.

薄膜分离法可去除并回收天然气中的重烃，方法简单且成本低。

hydrolyze *v.* 水解

例：COS and CS_2, although not corrosive in LPG, will hydrolyze slowly to H_2S in the presence of free water and cause the product to become corrosive.

氧硫化碳和二硫化碳在液化石油气中虽然没有腐蚀性，但会在自由水中缓慢水解成硫化氢，使产品具有腐蚀性。

hydrogen bond *np.* 氢键

例：Natural gas hydrates are created when certain small molecules, particularly methane, ethane, and propane, stabilize the hydrogen bonds within water to form a three-dimensional, cage-like structure with the gas molecule trapped within the cages.

天然气水合物是天然气中的小分子如甲烷、乙烷、丙烷，在水的氢键作用下形成的稳定三维笼状结构物质。

caustic wash *np.* 碱洗

例：When quantities of H_2S and CO_2 components are small, a simple caustic wash is both effective and economical.

当硫化氢和二氧化碳组分含量很低时，简单的碱洗既经济又有效。

amine treating *np.* 胺处理法

例：Amine treating is a very attractive alternative, especially when there is already an amine gas treating unit on site.

胺处理法是非常不错的备选方案，特别是现场有胺处理装置时更是如此。

amine solution *np.* 胺溶液

例：This is done by counter current washing with an amine, the H_2S being removed for sulphur recovery by heating the amine solution in a separate vessel thus regenerating the amine for recycle to the washing stage.

方法是，用胺溶液进行逆流洗涤，在另外的容器里加热该胺溶液去除硫化氢以回收硫，再次得到胺溶液用于循环逆流洗涤。

amine sweetening *np.* 胺法脱硫

例：<u>Amine sweetening</u> removes acid gases like hydrogen sulfide and carbon dioxide by subjecting the natural gas to amines and subsequently removing other trace hydrocarbons before the gas flows to the stripper.

胺法脱硫是指将天然气置于胺中，在气体流入汽提塔之前去除其他微量碳氢化合物，从而去除硫化氢（H_2S）和二氧化碳（CO_2）等酸性气体的天然气脱硫法。

overhead condenser *np.* 塔顶冷凝器

例：The stripper concentrates the methanol in the <u>overhead condenser</u> where it can be removed and further purified.

甲醇可在汽提塔顶部冷凝器里分离出来并进一步提纯。

atmospheric stripper *np.* 常压汽提塔

例：TEG is regenerated more easily to a concentration of 98%–99% in an <u>atmospheric stripper</u> because of its high boiling point and decomposition temperature.

由于沸点和分解温度较高，三甘醇在常压汽提塔中更容易达到98%～99%的浓度。

reactivator tower *np.* 再生塔

例：The treatment solution is circulated through an absorber tower and a <u>reactivator tower</u> in much the same way as the ethanolamine is circulated in the Girbotol process.

处理液在吸收塔和再生塔中循环，与乙醇胺法中的乙醇胺循环方式大致相同。

alkazid process *np.* 碱处理法

例：Other processes include the <u>alkazid process</u>, which removes hydrogen sulfide and carbon dioxide using concentrated aqueous solutions of amino acids.

其他方法包括碱处理法，即用高浓度氨基酸浓溶液去除硫化氢和二氧化碳。

hot potassium carbonate process *np.* 热碳酸钾处理法

例：The hot potassium carbonate process decreases the acid content of natural and refinery gas from as much as 50% to as low as 0.5% and operates in a unit similar to that used for amine treating.

热碳酸钾处理法可将天然气和炼厂气中的酸气含量从50%降至0.5%，且处理装置与胺处理装置类似。

molecular sieves *np.* 分子筛法

例：Molecular sieves is also an effective means of water removal and thus offer a process for the simultaneous dehydration and desulphurization of gas.

分子筛法也是一种行之有效的脱水手段，此方法可同时对气体进行脱水和脱硫。

amine process *np.* 胺法工艺

例：In general, batch and amine processes are used for over 90% of all onshore wellhead applications.

通常，陆上井口90%以上应用的是批量式处理和胺法工艺。

elemental sulfur *np.* 单质硫

例：Both reactions yield water and elemental sulfur.

这两个反应都能生成水和单质硫。

conversion process *np.* 转换工艺

例：Where large flow rates are encountered, it is more common to contact the produced gas stream with a chemical or physical solvent and use a direct conversion process on the acid gas liberated in the regeneration step.

当气体流量很大时，通常让采出气与化学或物理溶剂接触，并对再生过程中产出的酸气采用直接转换工艺处理。

liquid redox sulfur recovery process *np.* 液相氧化还原法

例：Liquid redox sulfur recovery processes are liquid-phase oxidation processes that use a dilute aqueous solution of iron or vanadium to remove H_2S selectively by chemical absorption from sour gas streams.

液相氧化还原法属于液相氧化法，使用含铁或钒的稀溶液，通过化学吸收选择性地去除酸气中的硫化氢。

Claus sulfur recovery process *np.* 克劳斯硫回收法

例：The Claus sulfur recovery process is the most widely used technology for recovering elemental sulfur from sour gas.
克劳斯硫回收法是从酸气中回收单质硫最常用的方法。

acid gas removal process *np.* 酸性气体脱除工艺

例：These processes can be used in place of an acid gas removal process.
在某些情况下，该工艺能取代酸性气体脱除工艺。

caustic soda solution *np.* 苛性钠溶液

例：For LPG and light gasolines therefore the mercaptans can be removed by counter current washing with caustic soda solution.
因此，在生产液化石油气和轻质汽油时，可以采用苛性钠溶液进行逆流洗涤以除去硫醇。

alkaline solutions *np.* 碱性溶液

例：Those mercaptans boiling below about 80 ℃ are readily dissolved in alkaline solutions but the solubility decreases rapidly above that temperature.
那些沸点在80℃以下的硫醇极易溶于碱性溶液，但温度一旦超过80℃，其溶解度就会迅速降低。

dehydration *n.* 脱水干燥

例：Because most of the water vapor must be removed from natural gas before it can be commercially marketed, all natural gas is subjected to a dehydration process.
天然气中含有大量水蒸气，因此在销售前必须进行脱水干燥。

dehydration plant *np.* 脱水装置

例：The permanent solution is removal of water prior to pipeline transportation, using a large offshore dehydration plant that is not often the most cost-effective solution.
管输前对天然气进行脱水处理是一劳永逸的方法，但用大型海上脱水装置通常并不是性价比最高。

process selectivity *np.* 流程选择

例：Process selectivity indicates the preference with which the process removes one acid gas component relative to (or in preference to) another.

流程选择是指选择一种能更好地脱去某种酸气组分的方法。

gas purification *np.* 天然气提纯

例：The processes that have been developed to accomplish gas purification vary from a simple once-through wash operation to complex multistep recycling systems.

天然气提纯的方法从简单的一次过洗作业发展到复杂多步循环系统。

batchtype process *np.* 批处理

例：The most common type of process for acid gas removal is the batchtype process and may involve a chemical process in which the acid gas reacts chemically with the cleaning agent, usually a metal oxide.

最常见的酸气脱离方法是批处理，批处理可能涉及化学过程，即酸气与清洗剂发生化学反应，清洗剂一般为金属氧化物。

solid-based media *np.* 固体介质

例：Solid-based media are typically nonregenerable, although some are partially regenerable, losing activity upon each regeneration cycle.

固体介质一般不可再生，即便有些部分可再生，但在每次再生循环中也会逐步失去活性。

dry sorption process *np.* 干式吸附工艺

例：Most dry sorption processes are governed by the reaction of a metal oxide with H_2S to form a metal sulfide compound.

大多数干式吸附工艺是通过金属氧化物与硫化氢反应，生成金属硫化物。

regenerable reaction *np.* 再生反应

例：For regenerable reactions, the metal sulfide compound can then react with oxygen to produce elemental sulfur and a regenerated metal oxide.

在再生反应中，金属硫化物与氧气反应生成单质硫和再生的金属氧化物。

contaminant/impurity *n.* 杂质

例：Hydrogen was increasingly used in hydrotreaters to help remove sulfur and other contaminants from various dirty streams and change the structure of some others.

氢越来越多地用于加氢处理装置，以脱除各种复杂油气流中的硫和其他杂质，并改变某些分子结构。

sales gas hydrocarbon dew point specification *np.* 销售用气态烃露点标准

例：The temperature to which the gas is cooled depends on whether it is necessary to meet a sales gas hydrocarbon dew point specification or whether substantial liquid recovery is desired.

气体冷却的温度取决于处理目的，销售用气态烃露点标准和大量回收液烃的温度标准不一致。

flare *v.* 火炬燃烧

例：Where there is no available gas pipeline, separated associated gas may be flared.

如果没有排气管道，分离出的伴生气可能会在火炬中烧掉。

reinjection into producing formations *np.* 回注产气层

例：Increasingly in such cases, separated gas is being conserved by compression and reinjection into producing formations for eventual recovery and sales.

在这种情况下，更多的分离气经压缩后回注产气层，以便后期开采及销售。

Section V

Wastewater Treatment
污水处理

oil refinery wastewater (ORW)　*np.* 炼油废水

例：The global demand for petrochemical and petroleum industry products unavoidably generates large volumes of oil refinery wastewater (ORW).

全球对石化和石油工业产品需求巨大，不可避免地会产生大量炼油废水。

wastewater treatment facility　*np.* 污水处理设备

例：The desalting process creates an oily desalter sludge and a high temperature salt water waste stream which is typically added to other process wastewaters for treatment in the refinery wastewater treatment facilities.

脱盐过程会产生含油污泥和高温脱盐污水，这些废物通常会并入其他工艺废水中，通过炼油厂的污水处理设备集中处理。

coagulation settling tank　*np.* 混凝沉降罐

例：With the development of oil mining method, the compositions of produced water become more complex, so the coagulation settling tank is the common equipment of pretreatment .

采油方法不断发展更新，油田采出水成分愈加复杂，混凝沉降罐成为预处理的常用设备。

walnut shell filter　*np.* 核桃壳压力滤罐

例：Walnut shell filter is the key equipment in oilfield water treatment system.

核桃壳压力滤罐是油田废水处理系统的关键设备。

oily sewage　*np.* 含油废水

例：Oily sewage is generated in the process of industrial production and oil extraction, mainly existing in the forms of dissolved oil, dispersed oil, floating oil and emulsified oil.

含油废水在工业生产和石油开采过程中产生，主要以溶解油、分散油、浮油和乳化油的形式存在。

floating oil *np.* 浮油

例：It was shown that great majority of oil is present in the form of floating oil in sewage water.

结果表明，废水中绝大多数油以浮油的形式存在。

deck drainage disposal *np.* 甲板污水排放

例：Disposal piles are particularly useful for deck drainage disposal.

排放管桩尤其适用于甲板污水排放。

scale buildup/scale deposition *np.* 结垢

例：Disposal piles contain no small passages subject to plugging by scale buildup.

排放管桩内没有任何小通道，不会发生结垢堵塞。

cooling water system *np.* 冷却水系统

例：Cooling water system is widely used in oil and gas industry to reject waste heat.

冷却水系统在油气工业中广泛应用，以进行余热回收。

flocculant *n.* 絮凝剂

例：In influent water and effluent wastewater treatment, clarification aids like coagulants and flocculants help remove suspended solids, including oil and organics.

在进水和排废水的处理过程中，混凝剂和絮凝剂等助滤剂有助于去除油和有机物等悬浮物。

dissolved oil *np.* 溶解油

例：Industrial wastewater containing oil not only pollutes the environment but also wastes resources, which is the key to treat with the emulsified and dissolved oil for purification of wastewater containing oil.

工业含油废水不仅会造成环境污染，还会导致资源浪费。去除乳化油和溶解油是净化含油废水的关键。

fiber ball filtration *np.* 纤维球滤料

例：Studies have shown that the Hydrophilic Fiber Ball Filtration is a high-efficiency and low-cost oil/water separation technology for highly emulsified oily wastewater treatment applications.

研究表明，亲水性纤维球滤料是一种效率高、成本低的油水分离技术，可用于处理高乳化含油废水。

salt contamination *np.* 盐侵污染

例：In most onshore locations the water cannot be disposed of on the surface, due to possible salt contamination, and must be injected into an acceptable disposal formation or disposed of by evaporation.

多数陆上油区污水不能直接地面排放，因为会使土壤受到盐侵污染，必须通过污水井注入适当地层或者在地面进行蒸发处理。

disposal well *np.* 污水井

例：The disposal well serves as a temporary collection of bilge water such as oily sewage produced in the engine room.

污水井可临时收集机舱中产生的含油污水等舱底水。

sedimentation basin *np.* 沉降池

例：In the process of petrochemical processing, a large number of industrial wastewaters will be produced, which can only be discharged after being treated by a sedimentation basin.

在石油化工生产过程中，会产生大量工业废水，须经过沉降池处理后才能排放。

anaerobic treatment *np.* 厌氧处理

例：Anaerobic treatment has many advantages over other biological method particularly when used to treat petroleum refinery wastewater.

与其他生物法相比，厌氧处理有许多优势，尤其是处理炼油厂废水。

discharge pipe *np.* 排泄管

例：Discharge pipe is a pipe through which fluids can be discharged.

排泄管是排泄流体的管道。

induced air flotation (IAF) device *np.* 诱导气浮装置

例：The industrial use of induced air flotation devices for oil wastewater separation began in 1969.

诱导气浮装置用于工业分离含油废水始于 1969 年。

reverse osmosis *np.* 反渗透法

例：One of the most reliable forms of water purification for offshore production is reverse osmosis.

反渗透法对海上采油水净化来说是最可靠的方式之一。

coarse graining *np.* 粗粒化

例：In the coarse graining method (one of the oil water treatment methods), lipophilic materials are used as the filler to make fine oil droplets condense on the hydrophobic surface.

粗粒化是一种处理含油废水的方法，这种方法采用亲油性材料作为填充物，使油粒附着在疏水性表面上。

oil skimmer *np.* 撇油器

例：An oil skimmer is a machine that removes floating oil and grease from liquid.

撇油器可去除液体中的浮油和油脂。

scale inhibitor squeeze *np.* 防垢剂挤注技术

例：Scale inhibitor squeeze is a type of inhibition treatment used to control or prevent scale deposition.

防垢剂挤注技术是一种控制或防止结垢的抑制措施。

sludge pumps *np.* 污泥泵

例：Sludge pumps are capable of moving viscous, sludgy materials often found in various industries including wastewater, manufacturing, and oil/gas production.

污泥泵可以搅动黏稠、淤泥状的物质，这些物质通常出现在废水处理、制造业和石油天然气生产等行业中。

fugitive emission *np.* 短时排放

例：The negative effects of fugitive emission compounds, such as volatile organic compounds and hazardous air pollutants, are not limited to global warming and the greenhouse effect.

短时排放化合物，如易挥发有机化合物和有害空气污染物等，所产生的负面影响不仅限于全球变暖和温室效应。

Section VI

Pipeline Transportation
管道运输

gas trunk line/gas transmission pipe line *np.* 输气干线

例：Gas trunk line is a pipeline designed for natural gas transmission from production areas to consumption points.

输气干线目的是将天然气从生产端输送到消费端。

crude trunk line *np.* 原油干线

例：Crude trunk lines move oil from producing areas to refineries for processing.

原油干线将油从油田输送到炼油厂进行加工。

crude oil pipeline *np.* 原油管道

例：The major long-distance pipelines are the crude oil pipelines running from the Gulf Coast to the Midwest.

主要长输管道是从墨西哥海岸到中西部的原油管道。

product pipeline *np.* 成品油管道

例：There are two types of oil pipeline: crude oil pipeline and product pipeline.

石油管道分为两种：原油管道和成品油管道。

pipeline oil/clean oil *np.* 管输油 / 净化油

例：Oil that is free of impurities to the extent that it will meet pipeline specifications is referred to as clean oil or pipeline oil.

原油中所含杂质数量在管输要求范围之内时，这种油被称为净化油或管输油。

grade *n.* 型号 / 级别

例：Products shipped include the several <u>grades</u> of gasoline, aviation gasoline, diesel, and home heating oils.

运输的成品油包括各种型号的汽油、航空汽油、柴油和家庭用燃料油。

batching *n.* 油品混输

例：<u>Batching</u> is done either with or without a physical barrier separating the two products.

油品混输可以用物理法隔开，也可直接接触进行顺序输送。

batching sphere *np.* 管输油品分隔球

例：A <u>batching sphere</u> can be inserted in the pipeline to form a physical barrier between batches of different products to maintain separation.

可以把管输油品隔离球作为物理屏障，放在两种油品之间，实现隔离。

natural gas (pipe) line *np.* 天然气管道

例：The most recent major pipeline to start operating in North America is a TransCanada <u>natural gas line</u> going north across the Niagara region, supplying 16 percent of all the natural gas used in Ontario.

北美地区最新投入使用的主要管道是横跨加拿大的天然气管道，向北穿过尼亚加拉地区，为安大略省提供 16% 的天然气。

onshore pipeline *np.* 陆上管道

例：<u>Onshore pipelines</u> provide economic, safe, and reliable transportation of crude oil, natural gas, and refined products.

陆上管道为石油、天然气及精炼产品提供经济、安全和可靠的运输。

buried pipeline *np.* 埋地管道

例：<u>Buried pipelines</u> should have protective coating applied in accordance with standards coating.

埋地管道应采用符合标准的防护涂层。

horizontal pipe *np.* 水平管道

例：Slug flow may even be encountered in <u>horizontal pipes</u> under steady-state conditions if the flow regime is not properly selected.

在稳定条件下，如果流动状态选择不当，段塞流也可能在水平管道中出现。

subsea line *np.* 海底管线

例：Subsea lines over 2000 miles have, until recently, been regarded as uneconomic because of the subsea terrain making pipeline installation and maintenance expensive.

直到现在，修建2000英里以上的海底管线都被视为不经济，因为海底地形复杂，管线铺设和维护相当昂贵。

offshore/marine/submarine pipeline *np.* 海底管道

例：Offshore pipelines convey oil and gas from subsea wells to the platform and from the seaboard platform for further process and shipping.

海底管道将油气从海底油井输送至钻井平台，并在沿海平台进一步加工和运输。

sea-going pipeline *np.* 海上可移动管道

例：The technology comprises large-diameter pipe structures manifolded together in tiers, essentially a sea-going pipeline.

这项技术需要大直径管汇叠置，实际上相当于海上可移动管道。

NGL transportation infrastructure *np.* 天然气液输送设施

例：The removal of natural gas liquids usually takes place in a relatively centralized processing plant, where the recovered NGLs are then treated to meet commercial specifications before moving into the NGL transportation infrastructure.

天然气液回收一般在比较集中的处理厂进行，回收的天然气液经过处理符合销售标准后，再进入天然气液输送设施。

LNG pipeline *np.* 液化天然气管道

例：The disadvantages of long-distance LNG pipelines stem from the fact that the gas must be kept at low temperature to maintain it in a liquid phase.

液化天然气长输管道的不利之处在于，气体必须维持在低温状态下才能保持其为液态。

partial loading *np.* 不满负荷运行

例：A LNG pipeline will also be harder to start up after a shut down than a vapor-phase pipeline or a crude oil pipeline; it would also be less suitable for operation at <u>partial loading</u>.

液化天然气管道与气体管道或原油管道相比，管道停输后启动更困难，而且不适于不满负荷运行。

high pressure gas pipeline *np.* 高压燃气管道

例：<u>High pressure gas pipeline</u> is defined as overland gas pipelines operating at pressure in the range of 100 to 200 bar.

高压燃气管道是指在10～20兆帕范围内工作的陆上天然气管道。

low pressure gas pipeline *np.* 低压燃气管道

例：Depending on a consumer category, there are different types of gas distribution pipelines: <u>low pressure gas pipelines</u> - for supplying gas to households, middle- and high-pressure ones - for conveying gas to industrial enterprises.

根据不同用户类别，气体分配管道也分为不同类型：为家庭供气的低压燃气管道和为工业企业供气的中高压燃气管道。

the gas supply line/the building line *np.* 室内燃气管道

例：<u>The gas supply line</u>, also known as the building line, is the plumbing that runs throughout the house.

室内燃气管道，也被称为建筑管道，贯穿整个房屋。

gas transmission and distribution system (UGTDS) *np.* 燃气输配系统

例：Urban <u>gas transmission and distribution system</u> mainly consists of three subsystems: storage station, gas network, and users.

城市燃气输配系统主要由三个子系统组成：储气站、燃气管网和用户。

gas pipeline network *np.* 燃气管网

例：At present, the urban <u>gas pipeline network</u> mainly uses electric resistance welding pipes, cast iron pipes and polyethylene pipes.

目前，城市燃气管网主要采用电阻焊管、铸铁管和聚乙烯管。

gas distribution network *np.* 配气管网

例：Gas distribution network is a system of external pipelines from a source to a gas consumer service line as well as facilities and engineering devices for them.

配气管网是从供气管网到用户服务管网的外部管道系统，包括其配套设施和工程设备。

high–pressure gas distribution piping system *np.* 高压配气管道系统

例：A high-pressure gas distribution piping system operates at a pressure higher than the standard service pressure delivered to the user.

高压配气管道系统的运行压力高于提供给用户的标准服务压力。

pipeline laying *np.* 管道敷设

例：The marine pipeline laying operation is developed by the construction method of land pipelines crossing rivers and lakes.

海上油气管道敷设作业由陆上管道穿越河流或湖泊的施工方法发展而来。

bottom tow method *np.* 底拖法

例：For the bottom tow method, the pipeline rests on the seabed, and a tug pulls it.

底拖法是将管道置于海床上，由拖船牵引。

reel–lay method *np.* 卷管式铺管法

例：The decisive advantage of the reel-lay method is that the pipes are welded onshore and thus independently of marine environmental influences, and pipeline production and pipe-laying can be carried out separately.

卷管式铺管法的绝对优势体现在管道焊接在陆地上进行，并不受海洋环境的影响，因此管道生产和管道铺设可分开进行。

J–lay method *np.* J 型铺管法

例：The J-lay method is a feasible way to install offshore pipelines in deep and ultra deep waters.

J 型铺管法适于深水和超深水海域铺管。

pipelaying vessel (PLV) *np.* 铺管船

例：Pipelaying vessel (PLV) is a specialized floating facility for submerged pipeline laying.

铺管船是铺设海底管道的专用浮式设备。

pipe fitting *np.* 管件

例：A major portion of the pipe fittings used in the oil and gas industries are made from carbon steel.

石油和天然气行业使用的大部分管件由碳钢制成。

pipeline diameter *np.* 管道直径

例：Transport of sales gas is done at high pressure in order to reduce pipeline diameter.

运输销售天然气的管道需要高压运行，以减小管道直径。

inner diameter *np.* 内径

例：Oil pipelines are made from steel or plastic tubes with inner diameter typically from 4 to 48 inches (100 to 1,220mm).

输油管道由钢或者塑料制成，其内径通常为 4～48 英寸（100～1220 毫米）。

reducer *n.* 减径管

例：A reducer is the component in a pipeline that reduces the pipe size from a larger to a smaller bore.

减径管是管道上缩小管道直径尺寸的部件。

block valve *np.* 截断阀

例：A block valve is used on gas transmission systems to isolate a segment of the main gas pipeline for inspection and maintenance, or for shutdown in the case of a natural disaster or pipeline damage.

截断阀用于输气系统，隔断某段主要输气管道以便进行检查和维护，或在发生自然灾害或管道损坏时进行管道封锁。

pipe tee *np.* 三通

例：Pipe tee is a type of pipe fitting which is T-shaped having two outlets, at 90° to the connection to the main line.

管道三通是一种 T 形管件，有两个出口与主管道直角连接。

external hot-water jacket *np.* 外加水套炉

例：The simplest is an <u>external hot-water jacket</u>, either for a pipe-in-pipe system or for a bundle.

对于双层套管或丛式管，最简单的方法是外加水套炉。

marine loading arm *np.* 输油臂

例：A <u>marine loading arm</u> is a device consisting of articulated steel pipes that connect a tankship such as an oil tanker or chemical tanker to a cargo terminal.

输油臂由铰接钢管组成，是油船或化学品运输船等与货运码头连接的介质。

single pipeline *np.* 单管道

例：By transporting multiphase well fluid in a <u>single pipeline</u>, separate pipelines and receiving facilities for separate phases, costing both money and space, are eliminated, which reduces capital expenditure.

用不同管线输送，用不同设备接收不同相态的流体，会耗费更多资金和空间，如果用单管道混相输送，就会大幅缩减成本。

multiphase transportation *np.* 混相输送

例：<u>Multiphase transportation</u> technology has become increasingly important for developing marginal fields.

混相输送技术对于海上边际油气田开发越来越重要。

multiphase pipeline *np.* 多相流管道

例：For offshore gas fields, where space is limited, the total production has to be transported via <u>multiphase pipelines</u> which may contain a three-phase mixture of hydrocarbon condensate, water, and natural gas.

海上气田由于空间狭小，所有产出物经多相流管道流出，其中的流体可能是凝析油、水和天然气的三相混合物。

three-phase mixture *np.* 三相混合物

例：These lines lie at the bottom of the ocean in horizontal and near-horizontal positions and may contain a <u>three-phase mixture</u> of hydrocarbon condensate, water, and natural gas flowing through them.

这些管线位于大洋底部，平行于或近平行于海底，其中的流体可能是凝析油、水和天然气的三相混合物。

multi-product pipeline *np.* 混油输送管道

例：Multi-product pipelines are used to transport two or more different products in sequence in the same pipeline.

混油输送管道用于在同一管道中按顺序输送两种或两种以上不同的油品。

transmit *n.* 混油

例：Some mixing of adjacent products occurs, producing interface, also known in the industry as "transmit".

临近油品会出现一些混合，从而产生界面，在该行业中这种现象也称为"混油"。

absorption rate *np.* 吸收率

例：At the receiving facilities this interface is usually absorbed in one of the products based on the absorption rates.

在油品的接收设备区，该界面通常会被其中一种油品吸收，这取决于吸收率。

commingled product *np.* 混合产品

例：Alternately, transmit maybe diverted and shipped to facilities for separation of the commingled products.

或者，混油会被转送或运输到用于分离混合产品的设备。

condensation *n.* 凝析

例：Pipelines may operate at very high pressures (above 1000 psig) to keep the gas in the dense phase thus preventing condensation and two-phase flow.

管道可能在高压下（每平方英寸1000磅以上）运行，以保持气体处于致密相，从而阻止发生凝析和两相流动。

two-phase flow *np.* 两相流动

例：Condensation subjects the raw gas transmission pipeline to two-phase, gas/condensate, flow transport.

天然气在管道内的流动为两相流动（天然气/凝析油）。

hydrocarbon condensate *np.* 凝析烃

例：Hydrocarbon condensate recovered from natural gas may be shipped without further processing but is typically stabilized to produce a safe transportable liquid.

从天然气中回收的凝析烃不再处理也可以运输，但一般需要进行稳定处理以符合安全运输要求。

liquid condensation *np.* 凝析液

例：Liquid condensation in pipelines commonly occurs because of the multicomponent nature of the transmitted natural gas and its associated phase behavior to the inevitable temperature and pressure changes that occur along the pipeline.

天然气组分繁多，在管道运输中会因温度和压力的变化而发生相态变化，出现凝析液。

natural gasoline *np.* 天然汽油

例：Natural gasoline (condensate) and LPG streams are often contaminated with acidic compounds.

天然汽油（凝析液）和液化石油气经常受酸性化合物污染。

pipeline drip *np.* 分液器

例：Pipeline drips installed near wellheads and at strategic locations along gathering and trunk lines will eliminate most of the free water lifted from the wells in the gas stream.

井口附近、集气管线及主要干线的重要位置处装有分液器，用来脱除井中气流带出的大部分游离水。

multiphase flow simulation package *np.* 多相流模拟程序

例：An appropriate multiphase flow simulation package must be used to calculate some of the unknown necessary variables, which are required for injection systems design.

目前注入系统需要设计一个合理的多相流模拟程序，来计算这些未知的必要参数。

stratified flow regime *np.* 层流流态

例：Gas pipelines have typically used slug catchers to dissipate the energy of the liquid slugs, to minimize turbulence, to ensure that the gas and liquid flow rates are low enough so that the stratified flow regime and subsequently gravity segregation can occur.

气体管道中使用段塞捕集器来分散液体段塞的能量，减少紊流，降低气体和液体的流速，保持层流流态，并产生重力分离。

flow regime *np.* 流动形态

例：This limitation is caused by the rheological properties of suspensions with high solid fraction and may depend on flow regime conditions.

这种局限性由流体中悬浮的大量固体碎屑的流变学性质所导致，也可能与流体的流动形态有关。

hydraulic pipeline study *np.* 管道水动力分析

例：In many applications, multiple operating points are available based on hydraulic pipeline studies or reservoir studies.

在很多情况下，通过管道水动力分析和油气藏研究，可以得到多个工况点。

insulated chamber *np.* 绝热容器

例：To maintain temperature, the pipe structures are contained within a nitrogen-filled, insulated chamber.

为了维持一定温度，这些管线都封存在充满氮气的绝热容器内。

network of insulated pipes *np.* 绝热管网

例：District heating or the teleheating system uses a network of insulated pipes through which heated water, pressurized hot water and sometimes steams are transported to the customers.

本地供热或者远程供热系统通过绝热管网为用户输送热水和加压热水，有时也输送热蒸汽。

chiller *n.* 冷却机

例：Suitable compressors and chillers are needed, but would be much less expensive than a natural gas liquefier and would be standard so that costs could be further minimized.

此外，还需要压缩机与冷却机，但这些普通的设备要比液化机便宜，以及规格标准，能进一步降低成本。

electrical resistance heating system *np.* 电阻加热系统

例：An <u>electrical resistance heating system</u> may be desirable for long offset systems, where available insulation is insufficient, or for shut-in conditions.

隔热性能不好，或者封闭情况下的长距离输送管线，使用电阻加热系统可能会更理想。

sulfur compound *np.* 含硫化合物

例：This odorization is provided by the addition of trace amounts of some organic <u>sulfur compounds</u> to the gas before it reaches the consumer.

在天然气中加入微量有机含硫化合物给天然气添味，之后送达客户端。

diamond lattice *np.* 金刚石型晶格

例：One of the two most common hydrate structures is a <u>diamond lattice</u>, formed by large molecules such as propane and isobutane.

天然气水合物有两种最常见结构，其中一种是金刚石型晶格，由丙烷或者异丁烷之类的大分子形成。

chemical inhibitor *np.* 化学抑制剂

例：<u>Chemical inhibitors</u> are injected at the wellhead and prevent hydrate formation by depressing the hydrate temperature below that of the pipeline operating temperature.

井口中注入化学抑制剂，使水合物温度低于管线运转温度，从而防止水合物生成。

hydrate inhibitor/hydrate depressant *np.* 水合物抑制剂

例：The development of alternative, cost-effective, and environmentally acceptable <u>hydrate inhibitors</u> is a technological challenge for the gas production industry.

可替代、低成本、环保型水合物抑制剂的研发，仍然是天然气工业的一个技术挑战。

low-dosage hydrate inhibitors (LDHIs) *np.* 低剂量水合物抑制剂

例：These new hydrate inhibitors, called "low-dosage hydrate inhibitors" (LDHIs), form the basis of a technique that does not operate by changing the thermodynamic conditions of the system.

这些新型水合物抑制剂称为"低剂量水合物抑制剂"，该技术可在不改变系统热动力学条件的情况下实施。

inhibitor injection system *np.* 抑制剂注入系统

例：Proper design of an inhibitor injection system is a complex task that involves optimum inhibitor selection, determination of the necessary injection rates, pump sizing, and pipeline diameters.

化学抑制剂注入系统的设计相当复杂，包括抑制剂优选、注入速度确定、泵型及管径选择等。

spent chemical *np.* 废弃化学药剂

例：A primary factor in the selection process is whether the spent chemical will be recovered, regenerated, and reinjected.

选择时需要考虑的首要因素是废弃化学药剂是否要回收、再生，以及循环注入。

surface-active chemical/surface active agent *np.* 表面活性剂

例：In contrast to other types of inhibitors, AAs, which are surface-active chemicals, do not prevent the formation of hydrate crystals but keep the particles small and well dispersed.

与其他类型的化学抑制剂不同，AAS 是一种表面活性剂，虽然不能抑制水合物晶体的形成，但是可以分散这些晶体小颗粒并防止其长大。

cryogenic condition *np.* 低温下

例：Because methanol has lower viscosity and lower surface tension, it makes for an effective separation from the gas phase at cryogenic conditions (below $-13\,^\circ F$) and is usually preferred.

由于甲醇的黏度和表面张力较低，很容易在低温下（低于 $-13\,^\circ F$）从气体中分离出来，因此更受人们青睐。

glycol recovery unit *np.* 乙二醇回收装置

例：Two new types of low-dosage inhibitors have been developed, which will enable the subsea gas transmission pipelines to handle increased gas volumes without additional glycol injection or extra glycol recovery units.

已经研发出两种新的、低剂量、适用于海底天然气集输管线的化学抑制剂，不需要再注入乙二醇和增设乙二醇回收装置就可以输送更多的天然气。

slug flow *np.* 段塞流

例：Water can condense in the pipeline, causing slug flow and possible erosion and corrosion.

水能在管线里凝结，造成段塞流和潜在的侵蚀及腐蚀。

water cut *np.* 含水率

例：These additives have limitations mainly in terms of water cut, where they require a continuous oil phase and therefore are only applicable at lower water cuts.

这些化学抑制剂主要在含水率方面有局限性。因为需要连续油相，所以这些化学抑制剂仅适用于含水率比较低的情况。

aqueous solution *np.* 水溶液

例：Natural, associated, or tail gas usually contains water, in liquid and/or vapor form, at source and/or as a result of sweetening with an aqueous solution.

天然气、伴生气和尾气中通常都含有以液相和气相形式存在的水，这些水主要来自脱硫过程中的水溶液。

gas dehydration capacity *np.* 天然气脱水能力

例：The inhibitor selection process often involves comparison of many factors, including capital/operating cost, physical properties, safety, corrosion inhibition, and gas dehydration capacity.

选择化学抑制剂通常需要对很多因素进行比较，包括基础建设和运行费、天然气物理性质、天然气脱水能力，以及安全、防腐等。

liquid desiccant (glycol) dehydration *np.* 液体干燥剂（乙二醇）脱水法

例：The most common of these are liquid desiccant (glycol) dehydration, solid desiccant dehydration, and refrigeration (i.e., cooling the gas).

最常见的方法有液体干燥剂（乙二醇）脱水法、固体干燥剂脱水法及冷却法（例如气体冷却）。

solid desiccant dehydration *np.* 固体干燥剂脱水法

例：The liquid desiccant dehydration and solid desiccant dehydration methods utilize mass transfer of the water molecule into a liquid solvent (glycol solution) or a crystalline structure (dry desiccant).

液体干燥剂和固体干燥剂脱水法是将大量水分子转移到液体溶剂（乙二醇）或结晶体（干燥剂）中。

absorption *n.* 吸收法

例：Among the different gas drying processes, absorption is the most common technique, where the water vapor in the gas stream becomes absorbed in a liquid solvent stream.

在不同的气体干燥方法中，吸收法是最常用的一种方法。例如液体溶剂可以用于吸收天然气中的水蒸气。

offloading *n.* 卸载

例：One concern is to avoid the potential for fires and explosions in the offloading of crude oil by pipes onto tankers for transport to a refinery.

把管道原油卸载到油轮并运至炼油厂时，要避免起火和爆炸的潜在风险。

vapor phase *np.* 气态

例：For land transportation, pipe lining natural gas in the vapor phase is still the preferred method.

对于陆路运输，天然气气态管道运输仍是首选的输送方法。

liquid phase *np.* 液态

例：Products being shipped must remain in a liquid phase rather than become a mixture of gas and liquid.

运输成品油时，油品必须保持液态，而不是气液混合状态。

Section VII

Pipeline Pressurization and Maintenance
管道增压与维护

filling station *np.* 加气站

例：The filling stations can be supplied by pipeline gas, but the compressors needed to get the gas to 3000 psig can be expensive to purchase, maintain, and operate.

加气站可以通过管道供应气，但将天然气加压到每平方英寸 3000 磅以上，所需压缩机的购置、维护及运行成本可能较高。

compressor station *np.* 压气站

例：One of the most important components of the natural gas transport system is the compressor station. These stations perform the essential task of compressing natural gas as it travels through pipelines.

压气站作为天然气运输系统中最重要的组成部分之一，其基本任务是压缩管道中的天然气。

natural gas centrifugal compressor *np.* 天然气离心式压缩机

例：Natural gas centrifugal compressor applications gas lift, down hole gas injection, landfill gas, natural gas gathering.

天然气离心式压缩机应用于气举、井下注气、填埋气、天然气集输。

axial flow compressor *np.* 轴流式压缩机

例：Axial flow compressors are a type of air compressors that move the air in a direction parallel with some axis.

轴流式压缩机是使空气沿平行于某一轴线方向运动的空气压缩机。

reciprocating compressor *np.* 往复式压缩机

例：Reciprocating compressors rely on an arrangement of inlet and outlet values combined with a piston and cylinder to compress the gas.

往复式压缩机依靠进口和出口阀组合以及活塞和气缸来压缩气体。

centrifugal compressor *np.* 离心式压缩机

例：Centrifugal compressors rely on a series of rotating impellers to supply centrifugal force to the gas, increasing its pressure.

离心式压缩机通过一系列旋转叶轮为气体提供离心力，增加压力。

compressor train *np.* 压缩机组

例：The two platforms each contain one compressor train comprising a four-stage compressor with a gas turbine driver.

这两个平台都有压缩机组，带有燃气轮机驱动的四级压缩机。

pulsation dampener *np.* 压力缓冲器

例：Pulsation dampeners have to be installed upstream and downstream of the compressor to avoid damages to other equipment.

压缩机的上游和下游都必须安装压力缓冲器，以避免对其他设备造成损坏。

pumping station *np.* 泵站

例：The oil is kept in motion by pump stations along the pipeline, and usually flows at speed of about 1 to 6 meters per second (3.3 to 19.7ft/s).

油品在管道中的流动速度为 1～6 米/秒，其动力来自管道沿线的泵站。

natural gas compression station *np.* 天然气压缩站

例：Strategic planning involves the determination of the shortest and most economical routes where they are built, the number of pumping stations and natural gas compression stations along the line, and terminal storage facilities, so that oil from almost any field can be shipped to any refinery on demand.

战略规划是确定最短及最经济的管道修建路线、沿线泵站和天然气压缩站数量，以及终端储存设施，以便任何一个炼油厂需要油时都可以从任何一个油田运送。

booster pump *np.* 增压泵

例：The design of booster pumps varies depending on the fluid and uses a single-stage compression mechanism that may be used to raise the pressure of gas that is already above ambient pressure.

增压泵的设计因流体而异，且采用单级压缩机原理，可将本已高于环周边压力的气体再升压。

boosting station *np.* 增压站

例：Natural gas compression is used to increase pipeline-operating pressures and the capability to transmit gas over long distances without additional boosting stations.

天然气压缩用于提高管道运行压力，在长输中不用再设增压站。

elevational change *np.* 高度变化

例：Slugs are normally formed from elevational changes in the inlet supply pipes, changes in gas supply flow rates, and changes in pressure and temperature during transmission.

段塞通常是由集气管道高度变化、气体流量变化、输送过程中温度和压力变化引起的。

undulating terrain *np.* 起伏地形

例：These pipelines vary in length between hundreds of feet to hundreds of miles, across undulating terrain with varying temperature conditions.

这些管道有的长数百英尺，有的长达数百英里，而且地势起伏不平，温度变化多端。

delivery pressure *np.* 输送压力

例：In some cases, little processing is needed; however, most natural gas requires processing equipment at the gas processing plant to remove impurities, water, and excess hydrocarbon liquid and to control delivery pressure.

在某些情况下，天然气几乎不需要处理；然而，大部分天然气还是要通过处理厂进行处理，以去除杂质、水和多余的烃液并控制外输压力。

pipeline-operating pressure *np.* 管线输送压力

例：In recent years, there has been a trend toward increasing pipeline-operating pressures.

近年来，管线输送压力有增加的趋势。

sales pipeline pressure *np.* 销售管道压力

例：If gas is produced at lower pressures than typical <u>sales pipeline pressure</u> (approximately 700–1000 psig), it is compressed to sales gas pressure.

如果天然气生产压力低于正常销售管道压力（700～1000psig），则要将其压缩到管道压力。

line capacity *np.* 管道输气量

例：Dehydration also increases <u>line capacity</u> marginally.

天然气脱水后还可增加管道输气量。

pipeline shipping charge *np.* 管道运输费

例：This new US supplied natural gas displaces the natural gas formerly shipped to Ontario from western Canada in Alberta and Manitoba, thus dropping the government regulated <u>pipeline shipping charges</u> because of the significantly shorter distance from gas source to consumers.

新天然气由美国供应，取代了此前从加拿大西部阿尔伯塔和马尼托巴省运送至安大略省的天然气。因此，政府管道运输监管费降低，因为从气源运至用户的距离显著缩短。

transmission loss *np.* 输送损失

例：The benefits of operating at higher pressures include the ability to transmit larger volumes of gas, lower <u>transmission losses</u> due to friction, and the capability to transmit gas over long distances without additional boosting stations.

高压输送天然气能增加天然气运量，减少因摩擦引起的输送损失，在长途输送中不用增设增压站。

cathodic corrosion protection *np.* 阴极防腐

例：The choice of high-grade steels as construction material, good insulation, <u>cathodic corrosion protection</u>, and continuous monitoring for leaks, including visual monitoring from aircraft or ground inspection, ensure a high level of safety.

管道采用具有良好绝缘性能的高级钢材，进行阴极防腐保护，并在高空或地面连续监测泄漏情况，保证管道高度安全。

internal corrosion *np.* 内壁腐蚀

例：Protecting pipelines from corrosion is achieved internally by the injection of inhibitors to mitigate <u>internal corrosion</u>.

注入缓蚀剂可以减轻内壁腐蚀，从而保护管道免受腐蚀。

external corrosion *np.* 外壁腐蚀

例：Although there are mitigation strategies for the <u>external corrosion</u> of oil and gas pipelines such as coating, lining, and cathodic protection, the pipeline failures are still frequent.

尽管涂层、衬里和阴极保护等策略可缓解油气管道外壁腐蚀，但管道故障仍然频发。

concrete coating *np.* 混凝土覆盖层

例：To protect pipes from impact, abrasion, and corrosion, a variety of methods are used, including wood lagging (wood slats), <u>concrete coating</u>, rock shield, high-density polyethylene, imported sand padding, and padding machines.

为防止管道受到撞击、磨损以及腐蚀，采用了各种措施，包括使用木制保护层（木板条加固）、混凝土覆盖层、石砌保护、高密度聚乙烯、进口沙子填充及填充机器等。

natural gas pipe marker *np.* 天然气管道标识

例：<u>Natural gas pipe marker</u> identifies critical gas line information at a glance by using bright safety colors and bold text.

天然气管道标识使用醒目的安全色和粗体文本，使主要输气管道信息一目了然。

precipitation of wax *np.* 结蜡过程

例：<u>Precipitation of wax</u> from petroleum fluids is considered to be a thermodynamic molecular saturation phenomenon.

油气流结蜡过程是热动力学分子饱和现象。

paraffin wax *np.* 石蜡

例：<u>Paraffin wax</u> molecules are initially dissolved in a chaotic molecular state in the fluid.

在液体中，石蜡分子起初的溶解状态没有规律。

onset of wax precipitation *np.* 析蜡点

例：This thermodynamic state is called the onset of wax precipitation or solidification.

这种热动力学状态称为析蜡点或者凝固点。

wax deposition *np.* 蜡沉积

例：In oil flow lines, wax deposition occurs by diffusion of wax molecules and crystals toward and attachment at the wall.

在原油流动管线中，蜡分子和晶体混合物分散并附着在井壁上形成蜡沉积。

wax crystal *np.* 蜡晶体

例：The wax crystals reduce the effective cross-sectional area of the pipe and increase the pipeline roughness, which results in an increase in pressure drop.

蜡晶体减小了管线的有效输送面积，增大了管壁粗糙度，从而导致压降增大。

wax slush *np.* 蜡泥

例：Slugging in this case would be beneficial in removing wax slush from the line.

这种情况下的堵塞对管线清除蜡泥有帮助。

pigging *n.* 清蜡

例：The effect is also an increase in production as there is no time lost by unnecessary depressurization, pigging, heating-medium circulation, or removal of hydrate blockage.

因为不用浪费时间去降压、清蜡、循环传热介质、解除水合物堵塞等，生产效率也提高了。

wax crystal modifier *np.* 蜡晶体改变剂

例：As the name implies, wax crystal modifiers are very effective at suppressing the pour points of crude oils because they suppress the wax crystal growth, thus minimizing the strength of their interactions.

蜡晶体改变剂能够抑制蜡晶体增长，有效降低原油倾点，从而降低相互作用强度。

wax dispersant *np.* 蜡分散剂

例：A <u>wax dispersant</u> may act to inhibit wax nucleation and change the type of wax crystals from plate or needle to mal or amorphous.

蜡分散剂能够抑制蜡成核，并改变蜡晶体的类型，将其从片状或针状变成非晶形。

wax chemical *np.* 清蜡剂

例：The selection of the <u>wax chemical</u> should be made after careful consideration of the produced hydrocarbon and facilities.

应该根据所生产的烃类和设备谨慎选择清蜡剂。

spherical pig *np.* 清管球

例：Many years ago <u>spherical pigs</u> were commonly used in natural gas pipelines to remove liquids – water and condensates.

多年前，天然气管道普遍使用清管球去除水和凝析油等液体。

cleaning pig *np.* 清管器

例：Pigging is an option, but only after very careful consideration of the system's performance to understand the dynamics of the <u>cleaning pig</u> and continuous removal of the wax cuttings ahead of it.

选择刮蜡之前必须仔细考虑系统性能，了解清管器工作的动力学特征，并知晓如何不断清除刮掉的蜡屑。

smart pig *np.* 智能清管器

例："<u>Smart pigs</u>" are used to detect anomalies in the pipe such as dents, metal loss caused by corrosion, cracking or other mechanical damage.

"智能清管器"用于检测管道中的异常情况，例如凹陷、腐蚀引发的金属损失，以及破裂或其他机械损伤。

pig-launcher station *np.* 清管器发射站

例：These devices are launched from <u>pig-launcher stations</u> and travel through the pipeline to be received at any other station down-stream, either cleaning wax deposits and material that may have accumulated inside the line or inspecting and recording the condition of the line.

这些清管设备由清管器发射站发出，穿过管道之后由下游的其他泵站接收，从而清理管线内部的蜡沉积，以及可能沉积在管道内的其他物质，或检查并记录管道内状况。

stuck pig *np.* **卡住的清管器**

例：Controlling the bypass of flow around the pig for such a purpose is difficult. Hence, many pigging operations end up in failure with stuck pigs.

为此控制流体绕过清管器比较困难。因此，很多清管器作业最后都因为清管器卡住而失败。

Section VIII

Oil Tank Transportation
油罐油轮运输

railway transportation *np.* 铁路运输

例：<u>Railway transportation</u> far behind pipelines as the second-most popular mode of transportation for oil, representing just 10 percent of oil transfer during its peak between 2010 and 2014.

作为第二大石油运输方式，石油铁路运输远远落后于石油管道运输，在 2010—2014 年的高峰时期，铁路运输仅占石油运输的 10%。

maritime transportation *np.* 海洋运输

例：Ocean transportation is fast and efficient long-distance delivery option, the containers and tanks used for the <u>maritime transportation</u> of oil also provide an additional security guarantee for the transported goods.

石油长途运输时，选择海洋运输更为快速、高效，海上石油运输使用的集装箱、油罐也为石油运输提供了额外的安全保障。

tank truck *np.* 油罐车

例：<u>Tank trucks</u> deliver gasoline to service stations and heating oil to houses.

油罐车把汽油运输至加油站或者把燃用油运送到用户家中。

railroad tank car *np.* 铁路油罐车

例：Canada will phase out old <u>railroad tank cars</u> in 3 years.

加拿大将在 3 年内淘汰老式铁路油罐车。

LPG tank trailer *np.* 液化石油气汽车罐车

例：<u>LPG tank trailers</u> are used to transport a variety of liquefied gases including liquid ammonia, propane, propylene and butadiene.

液化石油气汽车罐车用于运输各种液化气体，包括液氨、丙烷、丙烯和丁二烯。

LNG tank trailer *np.* 液化天然气罐车

例：<u>LNG tank trailers</u> are designed to transport liquified natural gas, and tank trailers are also widely across globe because they are cost-effective, strong and flexible.

液化天然气罐车用于运输液化天然气，由于其高性价比、坚固性和灵活性等优点，在全球广泛使用。

oil barge *np.* 油驳

例：<u>Oil barge</u> means a vessel which is not self-propelled and which is constructed or converted to carry oil as cargo in bulk.

油驳是指非自行推进的船舶，用于运输大宗油品。

tank barge *np.* 平底油船

例：The domestic <u>tank barge</u> industry is composed of approximately 4,000 barges and they account for the transport of millions of tons of cargo annually.

国内平底油船运输业拥有约 4000 只驳船，每年可运输几百万吨货物。

towing industry *np.* 拖运业

例：The petroleum industry, both offshore and onshore, has proved to be a stimulant to the <u>towing industry</u>.

事实证明，无论是海上还是陆上石油工业，均加速了拖运业的发展。

tug *n.* 拖船

例：The increasing demand for deeper drilling over the past twenty years has justified the building of larger and more powerful <u>tugs</u> and larger barges.

过去 20 年，深海钻井需求不断增加，发展更大、动力更强的拖船及驳船很有必要。

combination carrier *np.* 组合货船

例：There are various types of tankers: oil tanker, parcel tanker (chemical vessels), <u>combination carrier</u> (designed to carry oil or solid cargoes in bulk), and barges.

现今的轮船有多种：油轮、多隔舱零担油船（化学品船）、组合货船（用于大量运输原油或固体货物）及驳船。

LNG carrier *np.* 液化天然气船

例：An <u>LNG carrier</u> is a tank ship designed for transporting liquefied natural gas under the temperature of −162 °C.

液化天然气船用于运输温度在 −162℃以下的液化天然气。

tanker vessel *np.* 油轮

例：Today's <u>tanker vessels</u>, which include both ships and barges, are responsible for moving of the vast volumes of liquid cargoes.

目前的油轮包括两种：轮船和驳船，用于运输大量液态石油产品。

ocean tanker *np.* 远洋油轮

例：Significant volumes of natural gas are transported in the liquid phase as LNG, but these shipments are made by special <u>ocean tanker</u> rather than by long-distance pipeline.

大量天然气作为液化天然气以液态形式运输，通过专用远洋油轮，而不是长输管道。

LNG tanker *np.* 液化天然气油轮

例：To carry more gas in less space is one reason for the development of <u>LNG tankers</u>.

发展液化天然气油轮的原因之一是要在更小空间内运输更多天然气。

liquefaction plant *np.* 液化处理厂

例：Natural gas is moved from the producing fields in the gaseous form to a <u>liquefaction plant</u> near a shipping port where the gas is liquefied for loading aboard an LNG tanker.

天然气以气体形式从气田运往港口附近的液化处理厂液化，以便装载液化天然气油轮外运。

product carrier *np.* 成品油运输轮

例：The three most common categories are crude oil carriers, <u>product carriers</u>: which can carry clean (e.g., gasoline, jet fuel) and dirty (e.g. black oils): and parcel carriers (chemicals).

最常见的三种类型是原油运输轮、成品油运输轮和化学品运输轮，其中成品油运输轮可运输清洁产品（例如汽油、航空油）和非清洁产品（例如黑油）。

cargo tank *np.* 油轮油舱

例：Cargo tank is a tank intended primarily for the carriage of liquids, gases, solids, or semi-solids.

油轮油舱主要用于运输液体、气体、固体或半固体油气。

VLCC (Very Large Crude Carrier) *np.* 超级油轮

例：Crude carriers are classed as either VLCCs (Very Large Crude Carriers) or ULCCs (Ultra Large Crude Carriers) and are designed to transport vast quantities of crude oil over many long and heavily traveled sea routes.

原油油轮分为超级油轮和超大型（巨型）油轮，用于在许多繁忙长途航线上大量运输原油。

ULCC (Ultra Large Crude Carrier) *np.* 超大型油轮 / 巨型油轮

例：The development of ULCCs came about due to an array of factors, including Middle East hostilities that led to the closure of the Suez Canal, nationalization of oil fields in the Middle East and strong competition among international ship owners.

巨型油轮由于多种因素发展起来，有导致苏伊士运河关闭的中东战争，以及中东地区油田国有化，也有国际船东之间的激烈竞争。

loading port *np.* 装货港

例：New Brunswick will also refine some of this western Canadian crude and export some crude and refined oil to Europe from its deep water oil ULCC loading port.

新不伦瑞克省也将炼制运自加拿大西部的原油，并通过其深海巨型油轮装货港向欧洲出口部分原油和成品油。

shipping port *np.* 港口

例：Natural gas is moved from the producing fields in the gaseous form to a liquefaction plant near a shipping port where the gas is liquefied for loading aboard an LNG tanker.

天然气以气体形式从气田运至港口附近的液化处理厂，液化后装船外运。

single-point mooring *np.* 单点系泊

例：VLCCs and ULCCs typically load at offshore platforms or <u>single-point moorings</u> and discharge at designated lightering zones off the coast.

超级油轮和超大型油轮一般在海上平台或单点系泊处装载，然后在规定的海岸外过驳区卸载。

lightering *n.* 驳运

例："<u>Lightering</u>," offloading or transferring oil from large tankers to smaller ones, a process which can move 1,000 barrels per hour, is used so that the smaller vessels can enter smaller ports that the larger vessels cannot.

"驳运"即把原油从大油轮转运至小油轮，每小时可转运1000桶，以便小油轮进入大油轮无法通行的小港口。

LPG carrier *np.* 液化石油气船

例：Liquefied petroleum gas <u>(LPG) carriers</u> are specialized vessels that are used to transport liquefied natural gas (LNG) and liquefied petroleum gas (LPG) under controlled temperature and pressure.

液化石油气船是运输液化天然气和液化石油气的专用船舶，可以控制温度和压力。

carrying capacity *np.* 运载能力

例：As the trade in oil is the largest category internationally, <u>carrying capacity</u> allows the shipping industry to assess how many tankers are required and on which routes.

由于石油贸易在国际贸易中份额最大，因此航运业会根据运载能力估算所需油轮的数量及航线。

fuel gas receiving station *np.* 燃气接收站

例：Energas Technologies recently commissioned a <u>fuel gas receiving station</u> for a new 340 MW power plant in the coastal city of Tema in Ghana.

Energas Technologies公司最近委托建造了一座燃气接收站，用于加纳沿海城市特马的一座340兆瓦的新发电厂。

turbine flowmeter *np.* 涡轮流量计

例：The turbine flowmeter is often used to measure oil, organic liquid, inorganic liquid and cryogenic fluid, etc., and large crude oil pipeline stations also use it for trade settlement.

涡轮流量计常用于测量石油、有机液体、无机液体和低温流体等，大型原油输送管线站也用其进行贸易结算。

vortex street flowmeter *np.* 涡街流量计

例：Vortex street flowmeter is widely used in water supply systems of oil and gas plants.

涡街流量计广泛应用于油气加工厂供水系统。

velocity flowmeter *np.* 速度流量计

例：Velocity flowmeter is an instrument for measuring oil and gas flow in oil and gas pipeline.

速度流量计用于测量油气管道内的油气流速。

trucking *n.* 货车运输

例：In the US, 70% of crude oil and petroleum products are shipped by pipeline, 23% of oil shipments are on tankers and barges over water, trucking only accounts for 4% of shipments, and rail for a mere 3%.

在美国，70%的原油和石油产品通过管道运输，23%的石油通过油轮和驳船在水上运输，货车运输仅占4%，铁路运输仅占3%。

odorant *n.* 添味剂

例：Since natural gas as delivered to pipelines has practically no odor, the addition of an odorant is required by most regulations in order that the presence of the gas can be detected readily in case of accidents and leaks.

由于管路输送的天然气无色无味，所以大多数规定要求在天然气中加入添味剂，这样天然气一旦发生泄漏，就可以被立即发现。

near-adiabatic *adj.* 接近绝热的

例：Natural gas hydrates decompose very slowly at atmospheric pressure so that the hydrate can be transported by ship to market in simple containers insulated to near-adiabatic conditions.

天然气水合物在常压下分解十分缓慢，这样就可以用接近绝热的简易储罐海运至市场。

boil-off loss *np.* 蒸发逸失

例：It is difficult for liquefied natural gas to be stored for periods of time (months) without significant boil-off losses.

液化天然气很难在储存一段时间（几个月）后不大量蒸发逸失。

intermittent gas *np.* 间歇供气

例：Small volumes of intermittent gas are not economically attractive to the major gas sellers for liquefied natural gas facilities.

对于液化天然气设施的大型销售商来说，少量间歇供气在经济上没有吸引力。

payload capacity *np.* 负载能力

例：By careful control of temperature, more gas should be transported in any ship of a given payload capacity, subject to volume limitation and amount and weight of material of the pipe (pressure and safety considerations).

通过精心控制温度，根据货轮的负载能力（即容积限制、管线材料的压力和安全限制），尽可能多地输送气体。

moisture sorption capacity *np.* 吸湿能力

例：The moisture sorption capacity is not affected by variations in pressure, except where pressure may affect the other variables listed previously.

吸湿能力不受压力波动的影响，除非压力会影响到前面提到的其他因素。

static equilibrium capacity *np.* 静态平衡吸水量

例：Static equilibrium capacity is the water capacity of new, virgin desiccant as determined with no fluid flow.

静态平衡吸水量指没有流体流过时新干燥剂的吸水量。

dynamic equilibrium capacity *np.* 动态平衡吸水量

例：Dynamic equilibrium capacity is the water capacity of desiccant where the fluid is flowing through the desiccant at a commercial rate.

动态平衡吸水量指流体大量流过干燥剂时干燥剂的吸水量。

useful capacity *np.* 有效吸水量

例：Useful capacity refers to the design capacity that recognizes loss of desiccant capacity with time as determined by experience and economic consideration and the fact that all of the desiccant bed can never be fully utilized.

考虑吸水能力会随时间降低，且所有干燥剂都无法充分利用的情况下的吸水量是有效吸水量。

regasification *n.* 再汽化

例：Gas to solids involves three stages: production, transportation, and regasification.

天然气固化包括三个步骤：生产、运输和再汽化。

oil line *np.* 油路

例：Electrostatic oil line sensor (OLS) is one type of the electrostatic sensors based on electrostatic charging phenomenon and has been used in wear debris monitoring in oil lubricated tribological systems.

静电油路传感器（OLS）是一种基于静电电荷现象的静电传感器，已广泛应用于油润滑摩擦系统的磨损碎屑监测。

natural gas and liquid transmission company *np.* 天然气与液化气运输公司

例：The American Petroleum Institute collaborated with federal and state regulators, natural gas and liquid transmission companies, and local distribution companies.

美国石油协会与联邦和地方监管委员会、天然气与液化气运输公司以及当地的销售公司开展了合作。

Part V

Petroleum Refining
石油炼化

Section I

Primary Oil Refining
石油初炼

petroleum refining *np.* 石油炼制

例：Petroleum refining is the physical, thermal and chemical separation of crude oil into its major distillation fractions which are then further processed through a series of separation and conversion steps into finished petroleum products.

石油炼制指用物理、热力学和化学方法蒸馏原油，蒸馏后的主要馏分再进一步分离转化，最终形成石油产品。

refinery *n.* 炼油厂

例：If the pipeline has to be shut down, the production and receiving facilities and refinery often also have to be shut down because gas cannot be readily stored, except perhaps by increasing the pipeline pressure by some percentage.

由于气体不容易储存，除非适当地增加管线压力，否则管线出现故障关闭，气井、集气站和炼油厂也必须关闭。

primary refining *np.* 粗炼

例：Petroleum primary refining begins with physical separation, or the distillation, or fractionation, of crude oils into separate hydrocarbon groups.

石油粗炼是先对原油进行物理分离，即蒸馏或分馏，使原油分离成不同烃类。

physical separation *np.* 物理分离

例：The refining process is very complex and involves both chemical reactions and physical separations.

炼制过程非常复杂，既包括化学反应也包括物理分离。

Part V　Petroleum Refining　石油炼化

evaporation　*n.* 蒸发

例：<u>Evaporation</u> is conducted by vaporizing a portion of the solvent to produce a concentrated solution or thick liquor.

蒸发是使一部分溶剂汽化，生成浓缩溶液或黏稠液体。

drying　*n.* 干燥

例：Evaporation differs from <u>drying</u> in that residue is a liquid (sometimes a highly viscous one) rather than a solid.

蒸发和干燥的不同之处在于蒸发的残留物质是液体（有时是高黏度的液体），而不是固体。

distillation　*n.* 蒸馏

例：Evaporation differs from <u>distillation</u> in that the vapor is usually a single component, and even when the vapor is a mixture, no attempt is made in the evaporation step to separate the vapor into fractions.

蒸发和蒸馏的不同之处在于蒸发出来的蒸汽通常是单一组分，即使蒸汽是混合物，在蒸发阶段也不会将其分离成不同的馏分。

crystallization　*n.* 结晶

例：Evaporation differs from <u>crystallization</u> in that emphasis is placed on concentrating a solution rather than forming and building crystals.

蒸发和结晶的不同之处在于蒸发着重于将溶液浓缩，而不是生成和析出结晶。

condense　*v.* 冷凝

例：Because the hydrocarbons boil at different temperatures, their vapors must also <u>condense</u> at different temperatures.

因为各种烃类的沸点不同，所以其蒸汽的冷凝点也不尽相同。

basic fraction　*np.* 基本组分

例：The primary refining processes are the distillation of the feedstock into its <u>basic fractions</u>, and then the re-distillation of these, in separate towers, into highly concentrated intermediates.

粗炼工序是将进料蒸馏成为其基本组分，然后将这些组分在单独的炼塔中再蒸馏成为高浓度的半成品。

fraction/cut *n.* 馏分

例：When the hydrocarbons boil, their vapors condenses separately. So the oil breaks down into its <u>fractions</u> and forms different liquids.

当烃类沸腾时，其蒸汽分别冷凝。这样，原油就分离成各种馏分而形成不同的液体。

fractionating tower /fractionator *np.* 分馏塔

例：The oil is distilled in a very large steel tower, the technical name of which is the "<u>fractionating tower</u>".

原油在大钢塔内蒸馏，塔的专业名称是"分馏塔"。

fractionation plant *np.* 分馏厂

例：In some cases, a mixed stream of liquid hydrocarbons separated from natural gas at field processing plants is moved to a <u>fractionation plant</u> where the mixed stream is separated (fractionated) into individual products, including ethane, propane, and butane.

在某些情况下，还要把从天然气中分离出来的液体混合物输送到分馏厂，精馏后分离成不同的产品，包括乙烷、丙烷和丁烷。

atmospheric distillation *np.* 常压蒸馏

例：Fractions obtained from <u>atmospheric distillation</u> include naphtha, gasoline, kerosene, light fuel oil, diesel oils, gas oil, lube distillate, and heavy bottoms.

常压蒸馏得到的馏分包括：石脑油、汽油、煤油、轻燃料油、柴油、瓦斯油、润滑馏分和釜底重质油馏分。

vacuum distillation *np.* 减压蒸馏

例：In thermal cracking, heavy gas oils and residue from the <u>vacuum distillation</u> process are typically the feed.

在热裂解过程中，给料通常是重粗柴油和减压蒸馏后的渣油。

distilling units *np.* 蒸馏装置

例：<u>Distilling units</u> are the clever invention of process engineers who exploit the important characteristics of the distillation curve.

蒸馏装置是工艺工程师根据蒸馏曲线的重要特性而做出的巧妙发明。

distillation tower *np.* 蒸馏塔

例：Crude oil is separated in atmospheric and vacuum <u>distillation towers</u>, into "fractions" or "cuts" with various boiling-point ranges.

原油在常压和真空蒸馏塔中分离为不同沸程的馏分。

atmospheric distillation column *np.* 常压蒸馏塔

例：An <u>atmospheric distillation column</u> in a natural gas condensate unit usually consists of a single column with 3 pump-arounds and 2 side strippers.

天然气凝析装置常压蒸馏塔通常由一个单塔、三个泵转塔和两个侧汽提塔组成。

vacuum fractionator column/vacuum distillation tower *np.* 减压蒸馏塔

例：The injection of superheated steam at the base of the <u>vacuum fractionator column</u> further reduces the partial pressure of the hydrocarbons in the tower, facilitating vaporization and separation.

在减压蒸馏塔底部注入过热蒸汽是为了降低塔内碳氢化合物压力，加快汽化和分离。

A.V.distillation unit *np.* 常减压蒸馏装置

例：Crude oil atmospheric rectification tower is an important part of <u>A.V.distillation unit</u>.

原油常压精馏塔是常减压蒸馏装置的重要组成部分。

vacuum distillation column *np.* 真空蒸馏塔

例：The heavier fractions from the <u>vacuum distillation column</u> are processed downstream into more valuable products through either cracking or coking operations.

在裂化和焦化的下游过程处理下，真空蒸馏塔中的较重馏分可以加工成更有价值的产品。

rectification tower *np.* 精馏塔

例：The <u>rectification tower</u> is a kind of tower vapor-liquid contact device for rectification.

精馏塔是一种用于精馏的塔汽液接触装置。

vacuum flasher *np.* 真空闪蒸塔

例：They assume the vacuum flasher is part of the distillation unit.

他们认为真空闪蒸塔是蒸馏装置的一部分。

regenerator *n.* 再生器

例：The cat cracker comprises the reaction section, the regenerator and the fractionators.

催化裂化装置由三部分组成：反应器、再生器和分馏塔。

pre-heater *n.* 预热炉

例：Beside the first tower there is very large furnace, the "pre-heater," where the temperature of the oil is raised to 340 degrees Centigrade. The boiling oil and vapors then flow from the furnace into the fractionating tower.

第一个塔边有个非常大的炉子，即预热炉，原油在此加热到340℃后，沸腾的油和汽从预热炉流入分馏塔。

fractionation tray *np.* 塔盘

例：The desalted crude oil is then heated in a heat exchanger and furnace to about 750 degrees (°F) and fed to a vertical, distillation column at atmospheric pressure where most of the feed is vaporized and separated into its various fractions by condensing on 30 to 50 fractionation trays, each corresponding to a different condensation temperature.

脱盐原油通过换热器和高温加热炉加热至约750华氏度，进入常压立式原油分馏塔，在该塔中大部分原料都会汽化，并根据不同温度分别在30～50个塔盘上冷凝，每个塔盘对应不同的冷凝度，分流成不同馏分。

mist eliminator/mist extractor *np.* 除雾器

例：This hypothesis was confirmed by obtaining and inspecting samples of a vacuum distillation column mist eliminator.

通过对真空蒸馏塔除雾器进行取样和检测，这一假设得到证实。

vacuum flasher *np.* 闪蒸罐

例：In the vacuum flasher, straight run residue from the distilling unit, while it is still hot, is pumped to the flasher.

在真空闪蒸罐中，蒸馏装置产生的直馏渣油温度仍然较高时，会泵送至闪蒸罐中。

feed line (riser) *n.* 进料管

例：In the fluidized-bed process, oil and oil vapor preheated to 500 to 800 degrees (°F) is contacted with hot catalyst at about 1,300 (°F) either in the reactor itself or in the <u>feed line (riser)</u> to the reactor.

在流化床工艺中，原油和油蒸汽被预热至 500 ~ 800 华氏度，在反应器或者反应器进料管中与热催化剂在 1300 华氏度下接触。

charge pump *np.* 加料泵

例：The first piece of equipment important to the operation, the <u>charge pump</u>, moves the crude from the storage tank through the system.

加料泵能将原油从储油罐中泵出，是蒸馏操作中最重要的设备。

light hydrocarbon *np.* 轻烃

例：Less dense (lighter) crude oils (with higher API gravity) generally have a larger share of <u>light hydrocarbons</u>.

低密度（较轻）原油（API 重度较高）通常含有较多的轻烃。

high-boiling compound *np.* 高沸点成分

例：When cuts with boiling-points over 350 ℃ are required, it is necessary to do the distillation under the vacuum to reduce the temperature at which the material boils, as some <u>high-boiling compounds</u> start to decompose when heated to their boiling-point at atmospheric pressure.

要想取得沸点在 350℃ 以上的馏分，蒸馏必须在真空条件下进行，以便降低原料的沸腾温度。因为某些高沸点成分在大气压下加热到沸点时便开始分解。

multi-stage reactor *np.* 多级反应器

例：Depending on the products desired and the size of the unit, catalytic hydrocracking is conducted in either single stage or <u>multi-stage reactor</u> processes.

催化加氢裂化是在单级还是多级反应器中进行，要根据所需产品的类型及装置的大小来决定。

stripping tower *np.* 汽提塔

例：The side streams are each sent to a different small <u>stripping tower</u> containing four to ten trays with steam injected under the bottom tray.

侧流分别送至不同的小型汽提塔，该塔有 4～10 个塔盘，蒸汽从底部塔盘注入。

condenser *n.* 冷凝器

例：Potential source of emissions from distillation of crude oil are the combustion of fuels in the furnace and some light gases leaving the top of the <u>condensers</u> on the vacuum distillation column.

原油蒸馏过程中，加热炉会燃烧燃料，较轻气体则从减压蒸馏塔冷凝器顶部散出，这时可能产生气体排放。

denser (heavier) liquid *np.* 重质液体

例：When the crude liquid/vapor charge hits the inside of the distilling column gravity causes the <u>denser (heavier)</u> liquid to drop toward the column bottom, but the less dense(lighter)vapors start moving through the trays toward the top.

当原油气液混合物进入蒸馏塔中时，重力作用就使得重质液体组分流向塔底，轻质气体组分通过塔板到达塔顶。

trickle *n.* 滴，细流；*v.* 滴流，细细地流

例：The naphtha feed is pressurized, heated, and charged to the first reactor, where it <u>trickles</u> through the catalyst and out the bottom of the reactor.

石脑油进料经加压、加热后，进入第一个反应器，慢慢流过催化剂并从反应器底部流出。

boiling point *np.* 沸点

例：Historically, physical properties such as <u>boiling point</u>, density (gravity), odor, and viscosity have been used to classify oils.

历史上，油品分类一直是基于沸点、密度、气味和黏度等物理性质。

viscosity *n.* 黏度

例：Heavy fuel oil usually contains residuum that is mixed to a specified <u>viscosity</u>

with gas oils and fractionator bottoms.

重质燃料油通常含有与瓦斯油和分馏塔底油混合至一定黏度的渣油。

semisolid *n.* 半固态

例：Petroleum can include three phases: gaseous (natural gas), liquid (crude oil), and solid or <u>semisolid</u> (bitumens, asphalt, tars, and pitches).

石油包含三种形态：气态（天然气）、液态（原油）、固态或半固态（沥青、焦油）。

heavier compound *np.* 重组分

例：As the vapor bubbles cool a little, some of the hydrocarbons in them will change from the vapor to the liquid state; the temperature of the vapor drops and the lower temperature of the liquid causes whatever <u>heavier compounds</u> remain in the vapor to condense(liquefy).

当气泡温度降低时，其中的一些碳氢化合物就从气态变为液态；蒸汽温度降低，液体温度也降低，蒸汽中的重组分浓缩形成液体。

gas fractionation *np.* 气体分馏

例：PFAS removal was increased by 9 time by enhancement of <u>gas fractionation</u>.

气体分馏加强后可使 PFAS 的脱除率提高 9 倍。

side draw *np.* 侧馏分

例：So a <u>side draw</u> may be used to recirculate the liquid through a heater to drive off any lighter hydrocarbons.

因此，可能用侧馏分通过加热器再循环液体，排出轻烃。

straight run residue *np.* 直馏渣油

例：The <u>straight run residue</u> will crack if the temperature goes too high.

如果温度过高，直馏渣油将会裂解。

heavier fraction *np.* 重质馏分

例：<u>Heavier fractions</u>, which may not vaporize in the column, are further separated later by vacuum distillation.

在常压塔中，重质馏分可能无法汽化，需要采用真空蒸馏进一步分离。

atmospheric gas oil *np.* 常压瓦斯油

例：Thus, as the volatile constituents of the feedstock pass up the tower, fractionation is accomplished by a series of stages into fractions of various boiling ranges: (1) naphtha, (2) kerosene, and (3) <u>atmospheric gas oil</u>.

因此，当原料的挥发性成分上升到塔上部时，再经过几个阶段即可将其分馏成不同沸程的馏分：（1）石脑油；（2）煤油和（3）常压瓦斯油。

vacuum gas oil *np.* 减压瓦斯油

例：About the same amounts are distilled into the middle distillate and <u>vacuum gas oil</u> from conventional crude oils.

常规原油可蒸馏出约同等量的中间馏分和减压瓦斯油。

vacuum residuum *np.* 减压渣油

例：More naphtha is distilled from light crude oils and more <u>vacuum residuum</u> is obtained from heavy crude oils.

轻油中蒸馏出的石脑油量较多，而从重油中得到减压渣油量较多。

reflux *n.* 回流

例：<u>Reflux</u>, serving the same purpose as it does in a distilling column, is in the form of a liquid spray from the top of the vessel, using some of the cooled streams drawn off one of the trays.

回流与蒸馏塔的作用一样，将一部分从塔板中抽出的冷却液流从闪蒸罐的顶部向下喷洒。

vaporization *n.* 汽化

例：Current refinery technology is based on distillation, which typically accounts for over half of the total energy consumption due to <u>vaporization</u> (phase change) and the need for reflux.

目前的炼油技术以蒸馏为基础，由于汽化（相变）和回流的需要，蒸馏所需的能耗通常占总能耗的一半以上。

hydrocarbon vapors *np.* 烃蒸气

例：Steam is then injected into the full coke drum to remove <u>hydrocarbon vapors</u>, water is injected to cool the coke, and the coke is removed.

然后将水蒸气注入焦化完的焦炭塔中去除烃蒸气，注入水冷却焦炭，然后去除焦炭。

side stream *np.* 侧线

例：Within each atmospheric distillation tower, a number of <u>side streams</u> (at least four) of low-boiling point components are removed from the tower from different trays.

在每个常压蒸馏塔中，许多低沸点侧线馏分（至少 4 个）从不同塔盘上被去除。

distributor *n.* 分布器

例：The <u>distributor</u> catches the droplets, which drip down to the bottom.

分布器捕捉液滴，使其落到底部。

reboiler *n.* 再沸器

例：<u>Reboilers</u> are types of heat exchangers typically used to provide heat at the bottom of distillation columns.

再沸器是热交换器，常用于在蒸馏塔底部提供热量。

vacuum pump *np.* 真空泵

例：The <u>vacuum pump</u> at the top of the vessel maintains the low pressure in the vessel and continuously draws off any vapors that have not condensed and usually consist of small amounts of water and some hydrocarbon.

闪蒸罐顶部的真空泵保持罐内处于低压状态，不断抽出没有冷凝的气体，这些气体通常含有一些烃和少量水。

steam eductor *np.* 蒸汽喷射器

例：Some vacuum flashers use a device called a <u>steam eductor</u> to maintain the low pressure.

有些闪蒸罐用蒸汽喷射器来保持低压环境。

heat interchanger/heat exchanger *np.* 换热器

例：The desalted and dewatered crude oil is transported by pumps and passed through a series of <u>heat interchangers</u> to exchange heat with the hotter distillation products.

脱盐、脱水后的原油泵送过系列换热器，与温度较高的蒸馏品进行热交换。

compound column *np.* 复合塔

例：The compound column is actually a superposition of several simple distillation columns.

复合塔实际上是几个简单蒸馏塔组合在一起。

product blending *np.* 产品混合

例：Product blending is where the different petroleum fractions are combined together to make the final product.

产品混合是混合不同石油馏分，制成最终产品。

steam ejector *np.* 蒸汽喷射泵

例：Steam ejectors, or, more recently, vacuum pumps, are used to create a vacuum for evaporation of the light vacuum gas oil and heavy vacuum gas oil fractions.

蒸汽喷射泵，最近多称真空泵，用于产生真空以蒸发轻真空瓦斯油和重真空瓦斯油馏分。

barometric condenser *np.* 气压冷凝器

例：In most systems, the vacuum inside the fractionator is maintained with steam ejectors, vacuum pumps, and barometric condensers.

在大多数系统中，保持分馏塔内呈真空状态，要用蒸汽喷射泵、真空泵和气压冷凝器。

bubble cap *np.* 泡罩

例：The perforations in the trays are fitted with a device called bubble caps.

塔板孔眼上有一种叫作泡罩的构件。

furnace *n.* 加热炉

例：In atmospheric distillation, the desalted crude is heated in a heat exchanger and furnace to approximately 750°F then fed to a vertical distillation column under pressure.

在常压蒸馏中，脱盐原油在热交换器和熔炉中加热到大约 750 华氏度，然后在压力下进入立式蒸馏塔。

Part V Petroleum Refining 石油炼化

sour gas *np.* 酸性气体

例：The <u>sour gas</u> is sent to the refinery sour gas treatment system which separates the fuel gas so that it can be used as fuel in the refinery heating furnaces.

酸性气体经炼油厂酸性气体处理系统分离后，可作为炼油厂加热炉的燃料。

solvent recovery *np.* 溶剂采收

例：The <u>solvent recovery</u> process is based on the ability of certain compounds to dissolve certain classes of other compounds selectively.

溶剂采收的基本原理是某些化合物能够选择性地溶解其他某些种类的化合物。

solvent extraction *np.* 溶剂萃取

例：The most widespread application of <u>solvent extraction</u> is used in BTX recovery, especially for benzene.

溶剂萃取法最常用于采收混合芳烃时，特别是在采收苯时。

nonvolatile solute *np.* 非挥发性溶质

例：The objective of evaporation is to concentrate a solution consisting of a <u>nonvolatile solute</u> and a volatile solvent.

蒸发的目的是浓缩由非挥发性溶质和挥发性溶剂所组成的溶液。

extracting solvent *np.* 萃取用溶剂

例：Solvent extraction is the transfer of a solute species from its initial location to a solvent known as the <u>extracting solvent</u>.

溶剂萃取是将溶质类物质从其初始地转移到溶剂中，这种溶剂称为萃取用溶剂。

original solvent *np.* 原溶剂

例：The extracting solvent must be substantially immiscible with the <u>original solvent</u>.

萃取用溶剂在本质上不应和原溶剂相溶混。

liquid-liquid extraction *np.* 液—液萃取

例：When the solute is in solution the extraction process is called <u>liquid-liquid extraction</u>.

当溶质处于溶液之中，这种萃取过程被称为液—液萃取。

solid–liquid extraction *np.* 固—液萃取

例：If, on the other hand, the solute forms part of a solid (which need no be "dry") the process is termed <u>solid-liquid extraction</u>.

另一方面，如果溶质成为固体的一部分（不一定是"干"的），则这个过程被称为固—液萃取。

selective affinity *np.* 选择亲合性

例：In liquid-liquid extraction, the extracting solvent must have a suitably <u>selective affinity</u> for the appropriate solute which sometimes occurs in company with materials other than the original solvent.

在液—液萃取中，萃取用溶剂必须对适合的溶质具有适当的选择亲合性，因为这种溶质有时会与原溶剂以外的物质混在一起。

raffinate *n.* 萃余相

例：The rest of the hydrocarbon, which rises to the top, is called <u>raffinate</u>.

上升至萃取塔顶部的剩余烃类，被称为萃余相。

batch solvent processing *np.* 间歇式溶剂采收工艺

例：If the sulfolane is poured off, the aromatic compounds can be "sprung" from it by simple distillation. This two-step process is <u>batch solvent processing</u>.

将环丁砜倒掉，经简单蒸馏即可把溶解的芳烃分离出去。该两步法就是间歇式溶剂采收工艺。

water content *np.* 含水量

例：One of the most important advantages of DESs as a solvent instead of ionic liquids is the ability of DESs to customize physical properties and phase transitions by selecting the corresponding composition, relative composition, or <u>water content</u>.

DESs 代替离子液体作为溶剂有一个最大优点，即 DESs 能够通过选择相应的成分、组成或含水量来控制物理性质和相变。

physical solvent process *np.* 物理溶剂法

例：<u>Physical solvent processes</u> employ an organic solvent, low temperatures, or high pressure.

物理溶剂法在高压或低温下采用一种有机溶剂。

chemical solvent process *np.* 化学溶剂法

例：In chemical solvent processes, absorption of the acid gases is achieved mainly by use of alkaline solutions such as amines or carbonates.

化学溶剂法主要是用碱溶液如醇胺类或碳酸盐类来吸收酸气。

crystallizer *n.* 结晶器

例：Crystal size distribution (CSD) must be under control, it is a prime objective in the design and operation of crystallizers.

晶体粒度分布必须受到控制，这是结晶器设计和操作中的一个主要目的。

aromatic content test *np.* 芳香烃含量测试

例：Results of the two aromatic content tests may differ because of the way the two gas-chromatography techniques separate and identify aromatics.

由于两种气相色谱技术分离和识别芳烃的方式不同，两种芳香烃含量测试的结果可能不同。

BTX *np.* 混合芳烃

例：The most widespread application of solvent extraction is used in BTX recovery, especially for benzene.

溶剂萃取法应用于混合芳烃采收最为广泛，特别是用于苯采收。

aromatics concentrate *np.* 芳烃浓缩物

例：Benzene raffinate contains no benzene. It is the leftover of the aromatics concentrate after the goodies (benzene) are removed.

苯萃余物中并不含有苯，而是芳烃浓缩物脱除苯后的剩余物。

air preheater *np.* 空气预热器

例：Heat pipe air preheater is widely used in waste heat recovery of flue gas.

热管式空气预热器广泛应用于烟气余热回收。

vent *n.* 排气口

例：Air emissions arise from the process heater, vents, and fugitive emission.

空气排放物主要来自加热器、排气口和短时排放。

Section II

Secondary Oil Refining
石油精炼

secondary refining process *np.* 精炼

例：The <u>secondary refining processes</u> are designed to convert some of the distilled hydrocarbons into different molecular forms. Conversion processes can produce hydrocarbons which do not exist in reservoir crude.

精炼是将某些蒸馏后烃类转化为不同的分子结构。转化工艺可以产出原油中不曾有的烃类。

sweetening/desulfurization *n.* 脱硫

例：The secondary refining process for the treatment of toxic, corrosive and evil-smelling sulphur-compound impurities is known as "<u>sweetening</u>".

对这种有腐蚀性的、有毒的、气味极臭的硫化物杂质进行精炼，称为"脱硫"。

caustic washing *n.* 碱洗

例：<u>Caustic washing</u> and hydrodesulphurization are finishing processes designed to remove H_2S and mercaptan impurities.

碱洗和水洗脱硫是两种精炼工艺，用以脱去硫化氢（H_2S）和硫醇等杂质。

hydrodesulfurization *n.* 水洗脱硫 / 加氢脱硫

例：<u>Hydrodesulfurization</u> is a critical part of oil refining, in which sulfur compounds including sulfides, mercaptanes, and disulfides from petroleum fractions are removed by high-pressure hydrogen treatment.

加氢脱硫是石油精炼的关键步骤，通过高压氢处理去除石油馏分中的含硫化合物，包括硫化物、硫醇和二硫化物。

amine process *np.* 胺吸收法

例：Removal of acid gases, also called gas sweetening, is achieved primarily through one of two processes: the <u>amine process</u>, which removes both CO_2 and H_2S; or the iron oxide process, which removes H_2S selectively.

去除酸性气体，也称为气体脱硫，主要有两种实现方法：胺吸收法，同时去除 CO_2 和 H_2S；或氧化铁法，可以选择性地去除 H_2S。

sulfur–processing plant *np.* 硫处理装置

例：All <u>sulfur-processing plants</u> must be completely gastight: in the hazard zones, instruments and alarm devices are installed which automatically shut down the plants in case of danger.

硫处理装置需要具备高密闭性，即在高危区域，安装监测仪表和报警设备，一旦发生危险情况可以自动关闭装置。

clay refining *np.* 白土精制

例：By evaluating economic profit, hydrofining obviously prefer to <u>clay refining</u>.

通过经济效益评估，加氢精制工艺明显优于白土精制工艺。

chemical treatment *np.* 化学精制

例：The content of gum, total oxidation insoluble matters and olefins, etc, is decreased significantly by <u>chemical treatment</u> of coker diesel oil before hydro-refining.

对焦化柴油进行化学精制处理后，再加氢精炼，可显著降低焦化柴油中的胶质、总氧化不溶物和烯烃等的含量。

adsorption refining *np.* 吸附精制

例：The equipment of <u>adsorption refining</u> is simple and costing is low, but adsorption capacity is low also.

吸附精制的设备简单，成本较低，但受吸附剂的吸附容量限制。

wet flue gas desulfurization *np.* 湿法烟气脱硫

例：A zinc oxide enhanced by citric acid <u>wet flue gas desulfurization</u> process for the removal of medium-low concentration of SO_2 from non-ferrous metal smelting flue gas was proposed.

提出了一种柠檬酸强化氧化锌湿法烟气脱硫工艺，用于脱除有色金属冶炼烟气中的中、低浓度二氧化硫。

semi-dry flue gas desulfurization *np.* 半干法烟气脱硫

例：Computational fluid dynamics (CFD) combined with the two-fluid model (TFM) was used for simulation of water vaporization and semi-dry flue gas desulfurization process in a two dimensional powder-particle spouted bed (PPSB).

计算流体力学结合双流体模型相，对二维喷流床（PPSB）内水蒸气蒸发和半干法烟气脱硫过程进行数值模拟。

counter current washing *np.* 逆流洗涤

例：This is done by counter current washing with an amine, the H_2S being removed for sulphur recovery by heating the amine solution in a separate vessel thus regenerating the amine for recycle to the washing stage.

用胺溶液进行逆流洗涤，在另外容器里加热胺溶液去除硫化氢，回收硫，可再生胺溶液，用于循环洗涤。

caustic soda solution *np.* 苛性钠溶液

例：For LPG and light gasolines, the mercaptans can be removed by counter current washing with caustic soda solution.

生产液化石油气和轻质汽油，可以采用苛性钠溶液进行逆流洗涤以除去硫醇。

sour fuel system *np.* 含硫燃料系统

例：A certain amount of noncondensable light hydrocarbons and hydrogen sulfide pass through the condenser to a hot well, and then are discharged to the refinery sour fuel system or are vented to a process heater, flare or another control device to destroy hydrogen sulfide.

一些不能冷凝的轻烃和硫化氢通过冷凝器到达热水井，然后排放到炼油厂含硫燃料系统或加热器、火炬及其他控制设备以消除其中的硫化氢。

cracking *n.* 裂化

例：In cracking process, large hydrocarbon molecules are cracked or broken to form two or more smaller molecules.

在裂化工序中，大的烃分子裂解或分裂成两个或更多的小分子。

light cracking *np.* 轻裂解

例：To meet the higher and higher demand of gasoline in modern era, one after another cracking techniques have been invented, such as light cracking, thermal cracking, catalytic cracking and so on.

现代社会对汽油需求越来越多，裂解技术层出不穷，如轻裂解、热裂解和催化裂解等。

residual cut *np.* 底部组分

例：The main purpose of cracking is to increase the yield of lighter, more valuable fractions from medium and residual cuts.

裂解的主要目的是从中间和底部组分中生产出更多有价值的轻组分。

thermal cracking *np.* 热裂解

例：The main reaction in visbreaking is thermal cracking of heavy hydrocarbons, since resins are holding asphaltene and keep them attached to the oil.

减黏裂化的主要反应是重烃热裂解，因为树脂保持沥青质并使其附着在油中。

catalytic cracking *np.* 催化裂化

例：Catalytic cracking uses heat, pressure and a catalyst to break larger hydrocarbon molecules into smaller, lighter molecules.

催化裂化是在高温高压和催化剂的共同作用下，将较大的烃分子分解成更小、更轻分子的过程。

catalyst *n.* 催化剂

例：The catalyst is in a fine, granular form which, when mixed with the vapor, has many of the properties of a fluid.

催化剂呈细粒状，和水蒸气混合时具有很多流体特性。

molecular sieve catalyst *np.* 分子筛催化剂

例：Etherification activities of various molecular sieve catalysts were investigated with FCC light ends as feedstock carrying on a fixed bed micro reactor.

以炼油厂催化裂化汽油轻馏分为原料，在固定床微型反应器中对几种分子筛催化剂的醚化活性进行研究。

hot catalyst particle *np.* 热催化剂颗粒

例：The hot catalyst particle coming from the regenerator unit evaporate the feed gas oil upon contact in the riser, and the cracking starts as the gas oil vapors and the catalyst particles move upward in the reactor.

热催化剂颗粒从再生器进入反应器，与瓦斯油原料接触并使其蒸发，当瓦斯油蒸气和催化剂颗粒在反应器内向上移动时，裂化开始。

crystalline synthetic silica-alumina *np.* 合成结晶硅—铝

例：Most catalysts used in catalytic cracking consist of mixtures of crystalline synthetic silica-alumina, termed "zeolites," and amorphous synthetic silica-alumina.

催化裂化中使用的大部分催化剂，包括合成结晶硅—铝（又称为"沸石"），以及非晶合成硅—铝的混合物。

rare earth metal *np.* 稀土金属结晶

例：Most catalysts consist of a crystalline mixture of silica-alumina with small amounts of rare earth metals.

大多数催化剂是由硅—铝与少量稀土金属结晶混合制成的。

silica *n.* 硅

例：On the contrary, what's needed is an unusual catalyst, and in this case it's made of alumina, silica, platinum, and sometimes palladium.

恰恰相反，所需要的仅仅是一种由铝、硅、铂和钯构成的具有独特功能的催化剂。

platinum *n.* 铂

例：Catalysts commonly include platinum or palladium, which acts as a promoter.

催化剂通常包括铂或钯，它们起催化作用。

catalyst degradation *np.* 催化剂降解

例：In the platforming process, the first step is the preparation of the naphtha feed to remove impurities and reduce catalyst degradation.

在铂重整工艺中，第一步是制备石脑油原料，去除杂质，减少催化剂降解。

Part V　Petroleum Refining　石油炼化

catalytic cracking unit　*n.* 催化裂化装置

例：Oil refineries are responsible for 4%–6% of global CO_2 emissions, and 20%–35% of these emissions released from the regenerator of fluid catalytic cracking (FCC) units, which are the essential units for the conversion of heavier petroleum residues (vacuum gas oil) into more valuable products.

全球 4%～6% 的二氧化碳排放来自炼油厂，其中 20%～35% 来自流体催化裂化（FCC）装置。FCC 装置是将较重的石油残渣（真空瓦斯油）转化为更有价值产品的基本装置。

fixed-bed catalytic cracking reactor　*np.* 固定床催化裂化反应器

例：Catalytic hydrocracking normally utilizes a fixed-bed catalytic cracking reactor with cracking occurring under substantial pressure (1,200 to 2,000 psig) in the presence of hydrogen.

催化加氢裂化通常使用固定床催化裂化反应器，在有氢气且压力高（1200～2000psig）的条件下进行裂化。

fluidized-bed reactor　*np.* 流化床反应器

例：A number of different catalytic cracking designs are currently in use in the U.S., including fixed-bed reactors, moving-bed reactors, fluidized-bed reactors, and once-through units.

美国目前使用的一些催化裂化工艺有固定床反应器、移动床反应器、流化床反应器以及直流炉燃烧器。

fluid-bed catalytic cracking　*np.* 流化床催化裂化

例：Fluid-bed catalytic cracking became the most widely used process worldwide because of the improved thermal efficiency of the process and the high product selectivity achieved, particularly after the introduction of crystalline zeolites as catalysts in the 1960s.

流化床催化裂化由于工艺热效率提高，产品选择性高（特别是在 20 世纪 60 年代引入结晶沸石作为催化剂后），成为世界上应用最广泛的工艺。

moving-bed process　*np.* 移动床工艺

例：In the moving-bed process, oil is heated to up to 1,300 degrees (°F) and is passed under pressure through the reactor where it comes into contact with a catalyst flow in the form of beads or pellets.

在移动床工艺中，油加热到 1300 华氏度，在反应器中加压流过，与催化剂珠或颗粒流接触。

cracking tower *np.* 裂解塔

例：A casual passerby of a refinery can make an easy mistake by referring to the many tall columns inside as "cracking towers".

在炼油厂中，人们容易将很多高反应塔误认为"裂解塔"。

steam stripping *np.* 水蒸气汽提

例：The fluidized catalyst and the reacted hydrocarbon vapor separate mechanically in the reactor and any oil remaining on the catalyst is removed by steam stripping.

流化催化剂和已反应的烃蒸气在反应器中机械分离，然后水蒸气汽提将催化剂上的残余油带走。

hydrocracking *n.* 加氢裂化

例：Conversion processes include coking, hydrocracking, and catalytic cracking to break large molecules into smaller fractions; hydrotreating to reduce heteroatoms and aromatics, thereby creating environmentally acceptable products.

转化过程包括焦化、加氢裂化和催化裂化，将大分子分解为小分子；加氢处理以减少杂原子和芳香烃，从而创造出环保型产品。

hydrocracker *n.* 加氢裂化装置

例：This is in contrast to the gas from operations like distilling (and, as discussed later, the hydrotreater, hydrocracker, reformer, and others) where the gases contain only saturated compounds.

其他原油加工过程（如常压蒸馏、加氢处理、加氢裂化、催化重整等）产生的气体只含有饱和烃类，这些烃类的分离在饱和烃分离装置上进行。

hydrocracking feedstock *np.* 加氢裂化原料

例：Hydrocracking feedstocks are usually first hydrotreated to remove the hydrogen sulfide and ammonia that will poison the catalyst.

加氢裂化原料通常先加氢处理，以去除对催化剂不利的硫化氢和氨。

reactor *n.* 反应器

例: In conventional Claus plants with two <u>reactors</u> connected in series the conversion of the H$_2$S feed is ca. 95 %, and a third reactor connected in series increases the conversion to 96%.

在传统的克劳斯装置中,如有两个反应器,那么硫化氢的转化峰值是 95%,如果有三个反应器,峰值则可以达到 96%。

visbreaking *n.* 减黏裂化

例: The purpose of <u>visbreaking</u> is to reduce the viscosity of residual fuel oil, improve the pour point of oil, and maximize the production of distillate oil.

减黏裂化的目的是降低残渣燃料油黏度,改善油品的倾点,使馏分油产量最大化。

polymerization *n.* 聚合作用

例: Coke formation by <u>polymerization</u>, condensation, dehydrogenation and dealkylation, and further cracking will be the result of asphaltene and coke leaving the liquid phase (delayed coking).

通过聚合、缩聚、脱氢和脱烷基形成焦炭后,再进一步裂化使沥青质和焦炭离开液相(延迟焦化)。

paraffinic side chain breaking *np.* 石蜡侧链断裂

例: The possible reactions in visbreaking are: <u>paraffinic side chain breaking</u> which will also lower the pour point; cracking of naphthens rings at temperature above 482 °C (900 °F).

减黏裂化可能出现的反应有:石蜡侧链断裂,倾点降低;萘环在 482° C(900° F)以上的温度裂解。

hydrocarbon molecule *np.* 烃分子

例: Thermal cracking, or visbreaking, uses heat and pressure to break large <u>hydrocarbon molecules</u> into smaller, lighter molecules.

热裂解或减黏裂化是在较高温度与压力下,将较大的烃分子分解为更轻的小分子。

recycle ratio *np.* 回炼比

例：A new three-step anaerobic sequencing batch reactor (3S-ASBR) was developed to enhance methane productivity and energy yield by varying the reactor volumetric and recycle ratios.

开发了一种新型三步厌氧序批式反应器 (3S-ASBR)，通过改变反应器容积和回炼比提高甲烷产量和能量产率。

regenerator *n.* 再生器

例：The cat cracker comprises the reaction section, the regenerator and the fractionators.

催化裂化装置由三部分组成：反应器、再生器和分馏塔。

disengagement chamber *np.* 沉降器

例：In the older cat crackers, the disengagement chamber was called the reactor because that's where most of the chemical changes took place.

在老式的催化裂化装置中，沉降器称为反应器，因为大部分反应都在其中进行。

catalytic hydrocracking *np.* 催化加氢裂化

例：Catalytic hydrocracking normally utilizes a fixed-bed catalytic cracking reactor with cracking occurring under substantial pressure (1,200 to 2,000 psig) in the presence of hydrogen.

催化加氢裂化通常使用固定床催化裂化反应器，在有氢气且压力高（1200～2000psig）的条件下进行裂化。

catalytic dewaxing *np.* 催化脱蜡

例：Catalytic dewaxing is a hydrocracking process operated at elevated temperatures (280–400℃；536–752℉) and pressures, 2070–10,350kPa (300–500psi).

催化脱蜡是一种加氢裂化过程，在高温（280～400℃；536～752℉）和高压2070～10350kPa（300～500psi）下进行。

solid superacid catalyst *np.* 固体超强酸催化剂

例：Solid superacid catalysts have been proved to be very effective from the viewpoint of activity, selectivity and reusability in the preparation of various important chemicals.

从活性、选择性和可重复使用性的角度来看，固体超强酸催化剂在制备各种重要化学品方面非常有效。

counter current washing *np.* 逆流洗涤

例：For LPG and light gasolines, the mercaptans can be removed by <u>counter current washing</u> with caustic soda solution.

在生产液化石油气和轻质汽油时，可以采用苛性钠溶液进行逆流洗涤以除去硫醇。

coking *n.* 焦化

例：<u>Coking</u> is a cracking process used primarily to reduce refinery production of low-value residual fuel oils to transportation fuels, such as gasoline and diesel.

焦化就是用裂解方法将炼油厂产出的劣质重质燃料油转化为汽油、柴油等运输燃料。

parallel drum *np.* 并行塔

例：When the coke drum is filled with product, the feed is switched to an empty <u>parallel drum</u>.

当焦化塔填满时，进料就转移到一个空的并行塔中。

feed stream/feedstock *np.* 进料

例：In delayed coking operations, the same basic process as thermal cracking is used except <u>feed streams</u> are allowed to react longer without being cooled.

延迟焦化过程与热裂化基本一样，只是进料反应时间较长不用冷却。

heavy distillate *np.* 重馏分

例：Thermal cracking of <u>heavy distillates</u> for gasoline production is not selective and produces substantial quantities of gas and fuel oil together with the gasoline, which is also not of very good quality.

通过热裂化重馏分生产汽油不太适宜，因为同时会产生大量的气和燃料油，且汽油的质量也不太好。

sour gas treatment system *np.* 酸气处理系统

例：Hot vapors from the coke drums, containing cracked lighter hydrocarbon products, hydrogen sulfide, and ammonia, are fed back to the fractionator where they can be treated in the sour gas treatment system or drawn off as intermediate products.

从焦炭塔出来的热蒸汽含有裂解轻烃产品，例如硫化氢和氨，因此要返回到分馏塔经酸气处理系统处理，或作为中间产品去除。

process heater *np.* 加工热水器

例：Air emissions from coking operations include the process heater flue gas emissions, fugitive emissions and emissions that may arise from the removal of the coke from the coke drum.

焦化过程中的空气污染排放有：加热器烟气排放，短时排放，以及焦炭塔分离焦炭可能产生的排放。

coke drum *np.* 焦炭塔

例：The removal of coke from the drum can release particulate emissions and any remaining hydrocarbons to the atmosphere.

从焦炭塔去除焦炭会造成微粒排放，剩余的烃组分也会排放到大气中。

fluid coking *np.* 流体焦化

例：A number of different processes are used to produce coke; "delayed coking" is the most widely used today, but "fluid coking" is expected to be an important process in the future.

目前多种工艺可用于焦炭生产，"延迟焦化"使用最广，但"流化焦化"未来会成为重要工艺。

high pressure water jet *np.* 高压水射流

例：Typically, high pressure water jets are used to cut the coke from the drum.

一般通过高压水射流将焦炭从焦化塔中分离出来。

coke removal *np.* 焦炭分离

例：Wastewater is generated from the coke removal and cooling operations and from the steam injection.

废水是在焦炭分离、冷处理及蒸汽注入过程中产生的。

catalytic reforming *np.* 催化重整

例：<u>Catalytic reforming</u> could upgrade the quality of these naphthas, some with octane numbers in the range of 35 to 40, to as much as 90 octane, increasing the quality and quantity of the gasoline—making capacity at the same time.

催化重整可提高这些石脑油的质量，使其辛烷值达到 90（有些在 35 ~ 40 之间），提高汽油质量和产量，提高产能。

reformate cooler *np.* 重整反应冷却器

例：There are two potential corrosion problems in the <u>reformate cooler</u>: (1) from the cooling water and (2) due to the use of organic chloride to regenerate the catalyst.

重整反应冷却器面临两个潜在的腐蚀问题：一是冷却水带来腐蚀问题；二是使用有机氯化物再生催化剂带来腐蚀问题。

debutanizer *n.* 脱丁烷塔

例：The liquid product from the bottom of the separator is sent to a fractionator called a stabilizer, a column nothing more than a <u>debutanizer</u>.

将分离器底部分离出来的液态产物输送至稳定塔，即脱丁烷塔。

cat reformer *np.* 催化重整器

例：Refiners built new <u>cat reformers</u>, debottlenecked old ones, and introduced new catalysts into their existing ones to increase the amount of aromatics, which were high octane gasoline blending components.

炼油工程师研发了新的催化重整器，克服了旧工艺中的不足，加入新催化剂以提高汽油中芳烃（高辛烷值汽油调和组分）的含量。

dehydrocyclization *n.* 脱氢环化反应

例：The typical reactions in catalytic reforming are <u>dehydrocyclization</u>, isomerization, and dehydrogenation.

催化重整的反应通常有脱氢环化反应、异构化反应和脱氢反应。

multistage reforming *np.* 多级重整

例：A <u>multistage reforming</u> process comprises the passage of a refinery stream through at least two reforming stages in series .

多级重整工艺是指炼油厂流程至少经过两个系列重整阶段。

oligomerization *n.* 低聚 / 齐聚反应

例：Linear alkylates are produced either by separation from petroleum containing a mixture of linear and isomerized substances or by synthesis through ethylene <u>oligomerization</u>.

线型烷酸化物可以从含有线型和异构化物质的混合石油中分离出来，也可以通过乙烯齐聚反应合成。

nitration *n.* 硝化作用

例：The manufacture of chemicals from petroleum is based on the ready response of the various compounds types to basic chemical reactions, such as oxidation, halogenation, <u>nitration</u>, dehydrogenation, polymerization, and alkylation.

石油能生产化学品，主要是因为多种化合物的基本化学反应迅速，如氧化作用、卤化作用、硝化作用、脱氢作用、聚合作用和烷基作用。

alkylation *n.* 烷基化

例：<u>Alkylation</u> reaction of benzene and methanol produces desirable product p-xylene, thus understanding the structure-performance relationship in alkylation reaction is significant.

苯与甲醇的烷基化反应可得到理想的产物对二甲苯，因此了解烷基化反应中的结构 - 性能关系很重要。

isomerization *n.* 异构化

例：Skeletal <u>isomerization</u> of cyclohexane and butane over WO_3/ZrO_2 catalyst was studied to elucidate the reaction mechanisms.

为了阐明反应机理，研究了环己烷和丁烷在 WO_3/ZrO_2 催化剂上的骨架异构化。

exothermic reaction *np.* 放热反应

例：<u>Exothermic reaction</u> away from hot wall contributes to heat transfer intensification.

远离热壁的放热反应有助于强化传热。

pyrolysis reaction *np.* 热解反应

例：Gas-phase <u>pyrolysis reaction</u> with turbulent flow in helically coiled tubes is studied.

研究了螺旋管内紊流作用下的气相热解反应。

butene alkylation reaction *np.* 丁烯烷基化反应

例：The isobutane/butene alkylation reaction is one of the most crucial refining processes since it gives rise to high octane and high purity gasoline.

异丁烷 / 丁烯烷基化反应是生产高辛烷值、高纯度汽油的一项最关键工艺。

ON (octane number) *np.* 辛烷值

例：The higher the octane number is, the better gasoline resists detonation and the smoother the engine runs.

辛烷值越高，汽油抗爆性能越好，发动机运行越平稳。

RON (research octane number) *np.* 研究法辛烷值

例：Knock propensity is predicted by fuel antiknock metrics: antiknock index (AKI), research octane number (RON), and octane index (OI).

爆震倾向由燃料抗爆指数、研究法辛烷值和辛烷值指数来预测。

AKI (antiknock index) *np.* 抗爆指数

例：A total of seven fuels in a spark ignition engine under boosted operating conditions is investigated to determine whether knock propensity is predicted by fuel antiknock metrics: antiknock index (AKI).

对增压工况下火花点火发动机中的七种燃料进行了研究，以确定燃料抗爆指标（抗爆指数）是否可以预测爆震倾向。

motor octane number *n.* 马达法辛烷值

例：The average prediction error of motor octane number and research octane number were 0.2 and 0.18 respectively.

马达法辛烷值预测平均误差为 0.20；研究法辛烷值预测平均误差为 0.18。

Section III

Petroleum Fuels
石油燃料

petrochemical industry *np.* 石油化工行业

例：Catalytic reforming of straight run naphthas is a very important process for octane improvement and production of aromatic feedstocks for <u>petrochemical industries</u>.

直馏石脑油催化重整是石油化工行业辛烷值改进和芳烃原料生产的重要工艺。

petrochemical plant *np.* 石油化工厂

例：Another group of products pipelines is used to transport liquefied petroleum gases (LPG) and natural gas liquids (NGL) from processing plants in oil and gas-producing areas to refineries and <u>petrochemical plants</u>.

另一组成品油管道用于运输液化石油气和天然气凝析液，从油气田处理厂运至炼油厂和石油化工厂。

primary product *np.* 初级产物

例：The <u>primary products</u> of the refining industry fall into three major categories: fuels, finished nonfuel products, and chemical industry feedstocks.

炼化初级产品主要分为以下三大类：燃料油、非燃料产品及化工原料。

bitumen asphalt *n.* 沥青

例：Later <u>bitumen</u> and other heavy materials will be made from the residues.

随后，用这种渣油制取沥青和其他重质材料。

petroleum gas (refinery gas and LPG) *np.* 石油气（炼厂气或液化石油气）

例：In a simple refinery, the split would probably consist of these six basic cuts: <u>petroleum gas (refinery gas and LPG)</u>, gasoline, naphtha, kerosene, gas oil, residue.

在简易炼油厂中,蒸馏产品可能由下列六种基本组分构成:石油气(炼厂气或液化石油气)、汽油、石脑油、煤油、粗柴油和渣油。

fuel oil/fuel *np.* 燃料油 / 燃料

例:The phrase "fuel oil" is generally taken to imply a blend consisting essentially of the black oils obtained as residues from distillation or cracking, after the lighter hydrocarbon fractions have been removed.

"燃料油"通常是指以一种黑色残渣油为主的混合物,是原油去除轻烃组分后经蒸馏或裂解得到的。

residual fuel oil *np.* 残渣燃料油

例:A fuel oil that contains any amount of the residue from crude distillation or hydrocracking is a residual fuel oil.

原油蒸馏或加氢裂化后产生含有残渣的燃料油,这种燃料油称为残渣燃料油。

fuel–oil specification *np.* 燃料油规格

例:A requirement in almost all fuel-oil specifications is a limit on sulphur content.

几乎所有的燃料油规格中都有含硫量限制。

kerosene *n.* 煤油

例:The middle distillates refer to products from the middle boiling range of petroleum and include kerosene, diesel fuel, distillate fuel oil, and light gas oil.

中间馏分油是指处于石油沸点中间范围的产品,包括煤油、柴油燃料、馏分燃料油和轻瓦斯油。

lamp kerosene *np.* 照明煤油

例:In the 1860s, when modern refinery practices began, the main products from raw petroleum were lamp kerosene and residue for use as a lubricant.

19世纪60年代,现代炼油技术刚刚起步,从原油中提炼出的主要产品只有照明煤油和作润滑油的渣油。

diesel *n.* 柴油

例：Crude oil is sold in the form of gasoline, solvents, <u>diesel</u> and jet fuel, heating oil, lubricant oils, and asphalts.

原油以汽油、溶剂、柴油、航空燃油、加热原油、润滑油和沥青等形式出售。

residua (residuum 的复数) *n.* 渣油

例：<u>Residua</u>, ranging from atmospheric to vacuum residua, including those high in sulfur or nitrogen, can be used as the feedstock; and the catalyst can be either synthetic or natural.

无论常压渣油还是真空渣油，高硫渣油亦或高氮渣油，均可作为原料；催化剂既可以是合成的也可以是天然的。

gas oil *np.* 瓦斯油

例：The remainder of the crude oil includes the higher boiling lubricating oils, <u>gas oil</u>, and residuum (the nonvolatile fraction of the crude oil).

原油剩余物还有高沸点的润滑油、瓦斯油和渣油（原油的非挥发性成分）。

liquefied petroleum gas *np.* 液化石油气

例：The light refinery products, <u>liquefied petroleum gas</u> and gasoline, must be almost completely sulfur-free; for diesel fuels and light heating oils, a substantial sulfur reduction to 0.1%–0.5% is required by recent legislation.

轻质成品油、液化石油气及汽油必须实现无硫化；最新法规要求，柴油产品和轻质民用燃料油硫含量须大幅降至 0.1% ~ 0.5%。

heavy gas oil *np.* 重柴油

例：The cat cracked <u>heavy gas oil</u> can be used as feed to a hydrocracker or thermal cracker or as a residual fuel component.

催化裂化的重柴油可以作为加氢裂化或者热裂化的原料，也可用作残渣燃料油的调和组分。

reduced crude *np.* 拔顶油

例：Feedstocks to hydrocracking units are often those fractions that are the most difficult to crack and cannot be cracked effectively in catalytic cracking units.

These include: middle distillates, cycle oils, residual fuel oils and <u>reduced crudes</u>.
加氢裂化装置通常用于处理难以裂解及不能在催化裂化装置中有效裂解的产物，包括中间馏分油、循环油、残余燃料油和拔顶油。

light heating oil *np.* 轻质燃料油

例：The permissible sulfur limits for <u>light heating oil</u> in most countries are < 0.5 wt%, because of its use in domestic heating.

因为用于民用加热，大部分国家要求轻质燃料油的硫含量小于0.5%（质量分数）。

heavy fuel oil *np.* 重质燃料油

例：Both processes reduce the production of less valuable products such as <u>heavy fuel oil</u>, and increase the feed stock to the catalytic cracker and gasoline yields.

这两种方法都能降低价值较低产品的产量，如重质燃料油等，同时增加催化裂解的原料和汽油产量。

petroleum coke *np.* 石油焦

例：As part of the upgrading process, coking also produces <u>petroleum coke</u>, which is essentially solid carbon with varying amounts of impurities, and is used as a fuel for power plants if the sulfur content is low enough.

给产品提纯升级的焦化过程还会产出石油焦，主要是含各种杂质的固体炭，如果硫含量低，可用作发电厂燃料。

finished product *np.* 终端产品

例：Secondary refining processes are used to increase the production of higher-valued products such as gasoline and diesel fuel and improve the quality of the <u>finished products</u>.

石油精炼用于提高汽油和柴油等高价值产品的产量，并提高终端产品质量。

motor gasoline/motor spirit *np.* 车用汽油

例：About 90 percent of the petroleum products used in the U.S. are fuels with <u>motor gasoline</u> accounting for about 43 percent of the total.

美国90%的石油产品用作燃料，其中车用汽油占43%。

jet fuel *np.* 航空燃油

例：Crude oil is sold in the form of gasoline, solvents, diesel and <u>jet fuel</u>, heating oil, lubricant oils, and asphalts, or it is converted to petrochemical feedstocks such as ethylene, propylene, the butenes, butadiene, and isoprene.

原油以汽油、溶剂、柴油、航空燃油、加热原油、润滑油和沥青等形式出售，或者转化为石化原料出售，如乙烯、丙烯、丁烯、丁二烯和异戊二烯。

Avgas, Avtag, Avtur *n.* 航空汽油、航空涡轮汽油、航空涡轮煤油

例：The fuels that are exploded when vaporized with air to provide primary moving power include the aviation fuels (<u>Avgas, Avtag, Avtur</u>), industrial and domestic gases (LPG), motor spirits, diesel oil, and refinery gases.

有些燃料油与空气雾化后燃爆，可以提供原动力。这些燃料油包括航空油（航空汽油、航空涡轮汽油、航空涡轮煤油）、工业及民用液化气（LPG）、车用汽油、柴油和炼厂气。

higher octane product *np.* 高辛烷值产品

例：One could enrich the blend of gasoline with the addition of what would later be termed <u>higher octane product</u> obtained from selected premium crudes or "natural" or "casing head" gasoline yielded from highly saturated natural gas.

通过添加后来的所谓高辛烷值产品可以提升汽油性能，这些高辛烷值产品产自精选的高级原油和从高度饱和度天然气中生产出来的天然汽油。

gasoline with a higher octane *np.* 高辛烷值汽油

例：Catalytic cracking has largely replaced thermal cracking because it is able to produce more <u>gasoline with a higher octane</u> and less heavy fuel oils and light gases.

催化裂化工艺可以增加高辛烷值汽油产量，减少重质燃料油和轻气产量，因此广泛取代了热裂解技术。

syncrude/synthetic crude oil *n.* 合成原油

例：In GTL transport processes, the natural gas is converted to a liquid, such as <u>syncrude</u>, methanol and ammonia, and is transported as such.

天然气合成油（GTL）的运输是将天然气转化为液体，比如合成原油、甲醇

和氨，然后以液体形式运输。

syngas/synthetic gas *n.* 合成气

例：Methane is mixed with steam and converted to syngas or synthetic gas (mixtures of carbon monoxide and hydrogen) by one of a number of routes using suitable new catalyst technology.

甲烷混入蒸气，在适当的新催化剂作用下生成合成气（一氧化碳和氢气的混合物）。

Section IV

Chemical Raw Materials
化工原料

petroleum–chemical intermediate *np.* 石化半成品

例：Of the remaining 12%, just over half is refined into petroleum-chemical intermediates.

余下的 12% 的原油中，只有一半多一点炼制成石化半成品。

finished nonfuel product *np.* 非燃料产品

例：The primary products of the refining industry also include finished nonfuel products (solvents, lubricating oils, greases, petroleum wax, petroleum jelly, asphalt, and coke).

炼化初级产品还包括非燃料产品，如溶剂、润滑油、润滑脂、石油蜡、凡士林、沥青和焦炭。

solvent *n.* 溶剂

例：The synthetic organic chemicals industry supplies materials to a wide range of industrial operations, including plastics, fibers, and solvents, biochemical agents, food chemicals, and materials for construction.

有机合成化学品行业为许多工业提供生产原料，包括塑料、纤维、溶剂、生化药剂、食用化学品及建筑材料。

lubricating base stocks *np.* 润滑油基础油

例：A catalytic dewaxing process can be used to dewax a variety of lubricating base stocks; as such, it has the potential to replace solvent dewaxing, or even be used in combination with solvent dewaxing, as a means of relieving the bottlenecks which can, and often do, occur in solvent dewaxing facilities.

催化脱蜡工艺可用于各种润滑油基础油脱蜡，很有可能取代溶剂脱蜡，甚至可以与溶剂脱蜡联合使用，以减少溶剂脱蜡设备故障。

lubricating oil *np.* 润滑油

例：Lubricating oil is distinguished from other petroleum fractions by the high (>400℃) boiling point as well as their high viscosity.

润滑油与其他石油馏分不同，其沸点高（>400℃）、黏度高。

lighter lubricating oil *np.* 轻质润滑油

例：The wax which is removed from the lighter lubricating oil stocks has relatively large crystals and is known as "paraffin wax".

从轻质润滑油中除去的蜡的结晶体相对较大，俗称"石蜡"。

rolling oil *np.* 辊子油（用于金属加工的润滑油）

例：Typical classifications of lubricating oils include automotive lubricants, diesel-engine lubricants, steam-turbine oils, gear and transmission lubricants, metal-working lubricants (i.e. cutting oils), rolling oils (for metal-working) and compressor oils.

常用润滑油有机动车润滑油、汽轮机润滑油、齿轮及传动装置润滑油、金属加工（亦即切削油）润滑油、辊子油（用于金属加工的润滑油）及压缩机润滑油。

steam turbine oil *np.* 汽轮机油

例：T6001 is a multi-functional additive package suitable for premium R&O steam turbine oil.

T6001是多功能添加剂，适用于优质R&O汽轮机油。

lubricating grease *np.* 润滑脂

例：About 5% of the average barrel of crude is used in the production of a wide range of lubricating oils and greases, waxes, solvents, and asphalt for roads and weatherproofing.

每桶原油平均约5%用于生产各种润滑油、润滑脂、石蜡、溶剂，以及铺路和防水用的沥青。

petrolatum (petroleum jelly) *np.* 石蜡油（凡士林）

例：Wax constituents are solid at ordinary temperatures (25 ℃；77°F) whereas petrolatum (petroleum jelly) contains both solid and liquid hydrocarbons.

蜡组分在常温（25 摄氏度；77 华氏度）下是固体，而石蜡油（凡士林）中既有固体烃类也有液体烃类。

middle distillates *n.* 中间馏分油

例：Feedstocks to hydrocracking units are often those fractions that are the most difficult to crack and cannot be cracked effectively in catalytic cracking units. These include: <u>middle distillates</u>, cycle oils, residual fuel oils and reduced crudes.

加氢裂化装置常用于处理难以裂解及催化裂化装置中无法有效裂解的馏分，包括中间馏分油、循环油、残余燃料油和拔顶油。

motor oil/automotive lubricant *np.* 车用机油

例：There are two main groups of oils used in intermittent service, such as <u>motor</u> and aviation <u>oils</u>.

间歇性使用的油主要有两类，例如车用机油和航空机油。

graphite product *np.* 石墨产品

例：Coke also has nonfuel applications as a raw material for many carbon and <u>graphite products</u> including anodes for the production of aluminum, and furnace electrodes for the production of elemental phosphorus, titanium dioxide, calcium carbide and silicon carbide.

焦炭亦可作为非燃料，如用作许多碳和石墨产品的原料，生产铝电池的阳极，以及生产磷元素、钛白粉、碳化钙、碳化硅用的加热炉电极等。

paraffin wax *np.* 石油蜡

例：Petroleum waxes are of two general types: <u>paraffin wax</u> in distillates and microcrystalline wax in residua.

石油蜡一般分为两种：馏分油中的石蜡和残渣中的微晶蜡。

microcrystalline wax *np.* 微晶蜡

例：The wax which is removed from the heavier fractions has much smaller crystals and is usually referred to as "<u>microcrystalline wax</u>".

从重质润滑油中除去的蜡，其结晶体要小得多，通常称为"微晶蜡"。

intermediate paraffin *np.* 中间石蜡

例：On the other hand, <u>intermediate paraffin</u> distillates contain paraffin waxes and waxes intermediate in properties between paraffin and microcrystalline waxes.

中间石蜡馏分包括石蜡和性质介于石蜡与微晶蜡之间的蜡。

bitumen/asphalt *n.* 沥青

例：<u>Bitumen</u> is a dark viscous material and in fact is normally regarded as a solid at ambient temperature.

沥青是黑色黏性物质，常温下通常被视为一种固体。

harder asphalt *np.* 硬沥青

例：On the other hand, soft asphalts can be converted into <u>harder asphalts</u> by oxidation (air blowing).

另一方面，软沥青可以通过氧化（鼓风）转化为硬沥青。

petrochemical feedstock *np.* 石化原料

例：Crude oil is also converted to <u>petrochemical feedstocks</u> such as ethylene, propylene, the butenes, butadiene, and isoprene.

原油也可以转化为石化原料出售，如乙烯、丙烯、丁烯、丁二烯和异戊二烯。

methane *n.* 甲烷

例：Another product produced in atmospheric distillation, as well as many other refinery processes, is the light, noncondensible refinery fuel gas (mainly <u>methane</u> and ethane).

轻组分、非冷凝炼厂燃料气（主要是甲烷和乙烷）是常压蒸馏及其他炼油过程产出的另一种产品。

ethane *n.* 乙烷

例：The typical distillation cuts coming from atmospheric distillation and vacuum distillation are: Refinery gas - Made up of methane and <u>ethane</u>.

常压蒸馏和真空蒸馏的馏分通常是由甲烷和乙烷组成的炼厂气。

propane *n.* 丙烷

例：<u>Propane</u> is usually converted to propylene by thermal cracking, although some propylene is also available from refinery gas streams.

丙烷一般通过热裂化转化成丙烯，不过丙烯也可以从炼厂气中得到。

butane *n.* 丁烷

例：The lighter materials such as butane and naphtha are removed in the upper section of the tower and the heavier materials such as residual fuel oil are withdrawn from the lower section.

较轻的物质，如丁烷和石脑油，在塔的上部馏出，较重的残余燃料油从塔的下部取出。

isobutane *n.* 异丁烷

例：Zeolites with different topologies were investigated for isobutane alkylation.

不同拓扑结构的沸石在异丁烷烷基化反应中的作用得到研究。

olefin *n.* 烯烃

例：The cracking process results in the creation of olefins, so the C_4 and lighter stream not only contain methane, ethane, propane, and butanes but also hydrogen, ethylene, propylene, and butylenes.

裂解过程会产生烯烃，因此 C_4 和较轻蒸气不仅包含甲烷、乙烷、丙烷和丁烷，还包含氢、乙烯、丙烯和丁烯。

light olefin *np.* 轻烯烃

例：Efficient thermocatalytic transformation of CO_2 into light olefins, particularly ethene, is a challenge.

通过热催化，将二氧化碳有效转化为轻烯烃，特别是乙烯，极具有挑战性。

ethylene *n.* 乙烯

例：Established petrochemicals, such as ethylene and ammonia, were produced in larger and technically more advanced plants.

乙烯和氨等成熟石化产品，在规模更大和技术更先进的工厂进行生产。

propylene *n.* 丙烯

例：The process gives the highest yield of propylene, lighter olefins and aromatics for petrochemical operation from feedstocks which can include conventional FCC feeds and residual feeds.

在石化产品生产中，使用常规催化裂化原料和渣油原料，该工艺使丙烯、轻烯烃和芳烃的产量最大化。

butylene n. 丁烯

例：The various butylenes are more commonly obtained from refinery gas streams.

各种丁烯更多的是从炼厂气中获得的。

butadiene n. 丁二烯

例：Primary petrochemicals are methanol, ethylene, propylene, butadiene, benzene, toluene and xylene.

一次石化产品包括甲醇、乙烯、丙烯、丁二烯、苯、甲苯和二甲苯。

benzene n. 苯

例：Benzene extraction from HC-stocks does not solve this problem because of the cost and the question of reuse of the benzene recovered.

鉴于苯提取成本过高，以及苯再利用还存在问题，从碳氢化合物中提取苯难以实现。

toluene n. 甲苯

例：Benzene, toluene and xylene of aromatics are important basic raw materials for petrochemical industry.

芳烃中的苯、甲苯和二甲苯是石化工业的重要基础原料。

xylene n. 二甲苯

例：The aromatics are ringed hydrocarbons, such as benzene, toluene and xylene.

芳烃是环状碳氢化合物，如苯、甲苯和二甲苯。

methanol n. 甲醇

例：Petrochemical feedstocks can be classified into three general groups: olefins, aromatics, and methanol.

石油化工原料一般分为三类：烯烃、芳烃和甲醇。

ethanol n. 乙醇

例：Commonly used solvents are: toluene and xylene, ethanol, methanol, propyl alcohol, acetidin, chloroacetic, n-butyl acetate, and so on.

常用的溶剂有：甲苯、二甲苯、乙醇、异丙醇、甲醇、乙酸乙酯、乙酸甲酯、乙酸丁酯等。

aromatics n. 芳烃

例：Primary petrochemicals include olefins (ethylene, propylene and butadiene), aromatics (benzene, toluene, and the isomers of xylene) and methanol.

主要的石油化工制品包括烯烃（乙烯、丙烯、丁二烯）、芳烃（苯、甲苯和对二甲苯异构体）和甲醇。

aromatic hydrocarbon np. 芳香族烃

例：An aromatic hydrocarbon can be monocylic or polycylic. A monocylic aromatic contains one benzene ring with the configuration of six carbon atoms.

芳香族烃可以是单环芳烃或多环芳烃，单环芳烃只含一个由六个碳原子构成的苯环。

naphthene n. 环烷烃

例：The paraffins are converted to isoparaffins and naphthenes are converted to aromatics to capitalize on their higher octane numbers.

正构烷烃转变为异构烷烃，环烷烃转变为芳烃，都是为了有效提高辛烷值。

isomeric alkane np. 异构烷烃

例：Isomeric alkane can be obtained from long carbon chain isomeric mixed alkane.

从长碳链异构混合烷烃中可获得异构烷烃。

long-chain normal alkanes np. 长链正构烷烃

例：At increasing depth, and hence increasing temperature, the bitumen fraction loses features such as odd–even predominance in long-chain normal alkanes, optical activity, and the predominance of four- and five-ringed cycloalkanes.

深度增加，温度也上升，因此，沥青组分失去了其长链正构烷烃的奇偶优势和光学活性，同时也失去了四环和五环烷烃的优势。

normal paraffins np. 正链烷烃

例：Catalytic dewaxing processes based on selective hydrocracking of the normal paraffins use a molecular sieve-based catalyst in which the active hydrocracking sites are accessible only to the paraffin molecules.

对正链烷烃进行选择性加氢裂化后，催化脱蜡工艺通过使用分子筛催化剂，使得只有烷烃分子能发生有效加氢裂化。

triethylamine n. 三乙胺

例：Triethylamine (TEA) is a water-soluble flammable liquid tertiary amine $(C_2H_5)_3N$ that is used chiefly in synthesis (as of quaternary ammonium compounds)

三乙胺（TEA）是一种水溶性易燃液体叔胺（C_2H_5）$_3$N，主要用于合成（如季铵化合物）。

unsaturated hydrocarbon np. 非饱和烃

例：In fact, unsaturated hydrocarbons, which are not usually present in virgin petroleum, are nearly always manufactured as intermediates during the various refining sequences.

实际上，原油中一般不存在非饱和烃，非饱和烃几乎都是在各种炼制过程中作为中间产品产出的。

petroleum heavy end np. 石油重质尾部馏分

例：The various reactions of petroleum heavy ends, in particular the asphaltenes, indicate that these materials may be regarded as chemical entities and are able to participate in numerous chemical or physical conversions to, perhaps, more useful materials.

石油重质尾部馏分（尤其是沥青质）的各种反应，表明这些物质可以看作化学品，能通过各种化学和物理反应转变成更有价值的产品。

good-grade aromatic coke np. 高品质芳焦

例：The overall effect of these modifications is the production of materials that either afford good-grade aromatic cokes comparatively easily or the formation of products bearing functional groups that may be employed as a nonfuel material.

这些改进要获得的总效果是生产这样的材料：要么比较容易地产出高品质芳焦，要么形成功能族产品用作非燃料材料。

tetraethyl n. 四乙基

例：For decades, refiners have exploited the mysterious fact that the addition of tiny amounts of <u>tetraethyl</u> lead could substantially increase the gasoline octane.

添加极少量的四乙基铅可显著提高汽油的辛烷值，这是个神秘的事实，几十年来，炼油厂都是这么做的。

vinyl chloride np. 氯乙烯

例：A molecule of <u>vinyl chloride</u> is produced by cracking ethylene dichloride, which is a compound made by reaching ethylene and chlorine.

氯乙烯分子是由二氯乙烯裂解产生的，二氯乙烯是由乙烯和氯生成的化合物。

platinum n. 铂

例：There are commercial markets for the by-products of many refinery processes. Examples of these are pure sulphur, important in other areas of industry, and the <u>platinum</u> in some spent catalysts.

许多炼制过程中的副产品都有商业市场，如纯硫（在其他工业领域中十分重要），以及催化剂废料中的铂。

epoxy resin n. 环氧树脂

例：Benzene and cyclohexane are responsible for products such as nylon and polyester fibers, polystyrene, <u>epoxy resins</u>, phenolic resins, and polyurethanes.

苯和环己烷用来生产这样的产品，如尼龙和聚酯纤维、聚苯乙烯、环氧树脂、酚醛树脂和聚氨酯。

phenolic resin np. 酚醛树脂

例：<u>Phenolic resin</u> in alkali aqueous solution can be regarded as anionic polyelectrolyte and forms complex with diazo resin.

在碱性水溶液中，酚醛树脂被视为阴离子聚电解质，可与重氮树脂形成复合物。

polyester fiber np. 聚酯纤维

例：After weighing samples of hand split respectively, picking out of <u>polyester fiber</u> in paper bags and drying, weighing, calculating.

将称重后的试样分别进行手工拆分，拣出其中的聚酯纤维，放入纸袋内烘干、称重、计算。

naphtha n. 石脑油

例：Crude goes in, and the products go out—gases, gasoline, <u>naphtha</u>, kerosene, light gas oil, heavy gas oil, and residue.

原油进入蒸馏塔，得到气体、汽油、石脑油、煤油、轻柴油、重质油和残渣等产品。

desalted crude np. 脱盐原油

例：In atmospheric distillation, the <u>desalted crude</u> is heated in a heat exchanger and furnace to approximately 750°F then fed to a vertical distillation column under pressure.

在常压蒸馏中，脱盐原油在热交换器和熔炉中加热到大约750华氏度，然后加压进入立式蒸馏塔。

low-sulfur crude oil np. 低硫原油

例：Many countries have established a maximum sulfur content in fuels of 1 wt%–2 wt%. This value can be reached without additional treatment only with a few <u>low-sulfur crude oils</u>, whose supplies are limited.

许多国家都规定燃料油的最大含硫量为1%～2%。无须处理就能达到该标准的低硫原油只有几种，其供应量有限。

sweet crude np. 无硫原油

例：<u>Sweet crudes</u> have less than 1% sulfur by weight.

无硫原油的含硫量低于1%。

inorganic salt np. 无机盐

例：An <u>inorganic salt</u> is one that does not contain C-H bonds as opposed to an organic salt that contains C-H bonds.

无机盐不含碳—氢键，而有机盐则含有碳—氢键。

metal salt np. 金属盐

例：<u>Metal salt</u> is formed when a hydrogen atom is replaced by a metal ion.

金属离子取代氢原子后，就形成了金属盐。

Section V

Chemical Products
化工产品

petrochemical *n.* 石油化学产品

例：<u>Petrochemicals</u> are generally chemical compounds derived from petroleum either by direct manufacture or by indirect manufacture as by-products from the variety of processes that are used during the refining of petroleum.

石油化学产品一般指直接或间接从石油中提炼的化合物，这些化合物往往是石油炼制过程中的副产品。

synthetic materials (fibres, rubbers, plastics) *np.* 合成材料（纤维、橡胶、塑料）

例：These petroleum-chemical intermediates are used as feedstocks in the manufacture of <u>synthetic materials</u> (<u>fibres, rubbers, plastics</u>, etc.), fertilizers, insecticides, and even protein for animal feeds.

这些石化半成品可用来生产合成材料（纤维、橡胶、塑料等）、化肥、农药，甚至动物饲料蛋白。

chemical fertilizer *np.* 化肥

例：<u>Chemical fertilizers</u> have been widely used to achieve maximum productivity in conventional agricultural systems.

化肥已广泛用于传统农业，以实现产能最大化。

insecticide *n.* 杀虫剂/农药

例：<u>Insecticides</u> are widely used to control insect vectors and insect pests.

杀虫剂广泛用于防治病媒昆虫和害虫。

detergent *n.* 清洁剂

例：When added to a liquid, a <u>detergent</u> reduces its surface tension, thereby increasing its spreading and wetting properties.

加入液体时,清洁剂可以降低液体的表面张力,增加液体的扩散和润湿性能。

refrigerant *n.* 制冷剂

例: These petroleum products can be used as primary input to a vast number of products, including fertilizers, pesticides, paints, thinners, <u>refrigerants</u>, antifreeze, resins, sealants, insulations, latex, rubber compounds, and hard plastics.

石油产品可用作其他产品的主要原料,如化肥、农药、涂料、稀释剂、制冷剂、抗冻结剂、树脂、密封剂、绝缘材料、乳胶、橡胶化合物及硬质塑料等。

waterproofing agent *np.* 防水剂

例: Paraffin wax is also used for electrical insulation, in polishes, for the "defeathering" of poultry and as <u>waterproofing agent</u>.

石蜡还可以用于电绝缘,加入抛光剂中,用于家禽"脱毛",还可作防水剂。

binder *n.* 黏合剂

例: Because of its adhesive, plastic nature and waterproofing qualities, bitumen is widely used for road-making purposes, generally in the form of a <u>binder</u> for stone aggregates.

沥青具有黏性、可塑性及防水性,因此广泛用作筑路材料,通常用作碎石集料的黏合剂。

food carton *np.* 食品蜡纸盒

例: If the wax melted too readily, a stack of such <u>food cartons</u> would lead to stick together on a hot day.

如果蜡太易熔化,食品蜡纸盒在天气炎热时就会粘在一起。

Greek fire *np.* 希腊火药

例: <u>Greek fire</u> was a naphtha–asphalt mix; the naphtha provided the flame and the asphalt provided the adhesive properties that prolonged the incendiary effect.

希腊火药由石脑油和沥青混合而成,石脑油可以燃烧,沥青(或柏油)具有黏着性,可以延长燃烧效应。

PVC (polyvinyl chloride)　*np.* PVC 聚氯乙烯

例：In the catalytic conversion process known as polymerization, vinyl chloride becomes the well-known oil-based plastic, PVC (polyvinyl chloride).

经过所谓的"聚合反应"催化转换工艺，氯乙烯成为著名的油基塑料——PVC 聚氯乙烯。

celluloid　*n.* 赛璐珞

例：Celluloid is resulted from efforts to produce a synthetic horn.

赛璐珞的产生是因为试制合成角。

cellulose acetate　*np.* 醋酸纤维素

例：Cellulose acetate such as safety film and "cellophane", were the outcome of trying to reduce.

试图降低赛璐珞的易燃性促使了醋酸纤维素生成，如安全膜和"玻璃纸"。

artificial fibre　*np.* 人造纤维

例：Typical synthetic products that are foremost in the minds of most people nowadays are plastics and artificial fibres.

在当今大多数人心目中，塑料和人造纤维是典型的、最重要的合成品。

plastic article　*np.* 塑料制品

例：As may be gathered from this, the bulk of plastic articles is made by the process known as moulding.

由此可以得出结论，大多数塑料制品由模压工艺制造。

industrial plastic　*np.* 工业塑料

例：The wide range of industrial plastics can be broadly divided into two categories, the thermo-plastic and the thermo-setting products.

广泛使用的工业塑料大致分为两类：热塑性塑料和热固性塑料。

thermo–plastic material　*np.* 热塑性塑料

例：Thermo-plastic materials can be softened and re-softened repeatedly by the application of heat and pressure, provided they are not heated to such an extent that chemical decomposition takes place.

对热塑性塑料来说，只要其受热程度不致引起化学分解就可以通过加压加热反复软化。

thermo-setting plastics *np.* 热固性塑料

例：Thermo-setting plastics undergo chemical change when subjected to heat and pressure.

热固性塑料受压受热后会发生化学变化。

nitrogenous fertilizer *np.* 氮肥

例：Nitrogenous fertilizers are those fertilizers which contain N as the chief component in their final product.

氮肥是指最终产品中的主要成分是氮的肥料。

phosphate fertilizer *np.* 磷肥

例：Phosphate fertilizer is a fertilizer that is high in phosphorous.

磷肥是一种含磷量高的肥料。

high-molecular polymer *np.* 高分子聚合物

例：A polymer with the very low degree of polymerization (1–100) is called an oligomer, and only when the molecular weight is up to 10–106 (such as plastics, rubber, fibers, etc.), it is called a high-molecular polymer.

聚合度很低（1~100）的聚合物称为低聚物，只有当相对分子质量达到10~106（如塑料、橡胶、纤维等）时，才称为高分子聚合物。

herbicide *np.* 除草剂

例：Herbicide is usually a chemical agent used to kill or inhibit the growth of unwanted plants, such as residential or agricultural weeds and invasive species.

除草剂通常是一种化学药剂，用于杀死或抑制有害植物的生长，如住宅区或农业区的杂草及入侵物种。

synthetic resin *np.* 合成树脂

例：Synthetic resin, short for resin, is artificial synthesized high molecular polymer.

合成树脂，简称树脂，是人工合成的高分子聚合物。

oil varnish *np.* 油漆

例：Oil varnish is essentially a mixture of a drying oil and a resin.
油漆本质上是一种干燥油和树脂的混合物。

chemical reagent *np.* 化学试剂

例：Chemical reagents offer a wide variety of chemicals and solutions suited for laboratory testing of incoming raw materials, in process testing, and the testing of finished products.
化学试剂有多种化学品和溶液，适用于来料的实验室测试、过程测试和成品测试。

synthetic drugs *np.* 合成药品

例：Synthetic drugs are created using man-made chemicals rather than natural ingredients.
合成药品是由人造化学物质，而不是天然物质制造出来的。

daily chemical products *np.* 日用化学品

例：Daily chemical products refer to daily chemicals, which are scientific and technological chemicals commonly used by people on daily days, including shampoo.
日化产品是指日用化学品，是人们日常常用的科技化工产品，包括洗发水。

tackiness agent *np.* 胶黏剂

例：Tackiness agent - additive used to increase the adhesive properties of a lubricant, improve retention, and prevent dripping and splattering.
胶黏剂是一种添加剂，用于增加润滑剂的黏结性，提高固位力，防止滴落和飞溅。

primary explosives *np.* 起爆药

例：Primary explosives are substances that can rapidly change from deflagration to detonation and generate shock waves to detonate secondary explosives.
起爆药能迅速从爆燃变为爆轰并产生冲击波引爆二次炸药。

参考文献

[1] Arnold K, Stewart M. Surface Production Operations: Vol. 2: Design of Gas-Handling Systems and Facilities [M]. Houston: Gulf Professional Publishing, 1999.

[2] Majid Tolouei-Rad. Drilling Technology [M]. London: IntechOpen, 2021.

[3] Mokhatab S, Poe W A, Speight J G. Handbook of Natural Gas Transmission and Processing[M]. Waltham: Gulf Professional Publishing, 2006.

[4] Norman J. Hyne. Nontechnical Guide to Petroleum Geology, Exploration, Drilling, and Production. PennWell Cooperation [M]. Tulsa: Pennwell, 2012.

[5] Pena S A, Abdelsalam M G. Orbital Remote Sensing for Geological Mapping in Southern Tunisia: Implication for Oil and Gas Exploration [J]. Journal of African Earth Sciences, 2006.

[6] Tissot B P, Welte D H. Petroleum Formation and Occurrence: A New Approach to Oil and Gas Exploration [M]. Berlin: Springer-Verlag, 1978.

[7] 江淑娟. 石油科技英语 [M]. 北京：石油工业出版社，2007.

[8] 江淑娟，吴松林. 石油勘探英语 [M]. 北京：石油工业出版社，2002.

[9] 江淑娟，吴松林. 油田开发英语 [M]. 北京：石油工业出版社，2002.

[10] 江淑娟，吴松林. 石油钻井英语 [M]. 北京：石油工业出版社，2003.

[11] 江淑娟，吴松林. 石油化工英语 [M]. 北京：石油工业出版社，2002.

[12] 吴松林，江淑娟，杨国俊. 石油经济与管理英语 [M]. 北京：石油工业出版社，2003.

[13] 王琪，李冰. 石油开采 [M]. 北京：石油工业出版社，2014.

[14] 刘葆，王超，康国强. 油气销售 [M]. 北京：石油工业出版社，2015.

[15] 李滨，高苏彤. 油气地面处理 [M]. 北京：石油工业出版社，2014.

[16] 随岩，秦艳霞. 石油炼制 [M]. 北京：石油工业出版社，2014.

[17] 刘江红，李健，李超. 石油环境保护 [M]. 北京：石油工业出版社，2014.

[18] 王利国，孙彩光，刘典忠. 非常规油气开发 [M]. 北京：石油工业出版社，

2016.
- [19] 刘亚丽，徐国伟，吴冠乔. 天然气开采 [M]. 北京：石油工业出版社，2016.
- [20] 刘佰明，乔洪亮，徐国伟. 海外油气运营 [M]. 北京：石油工业出版社，2016.
- [21] 朱年红，刘立安，赵巍. 油气储运 [M]. 北京：石油工业出版社，2014.
- [22] 郭玉海，范金宏，张曦. 石油地质与勘探 [M]. 北京：石油工业出版社，2015.
- [23] 王红，谷秋菊，马焕喜. 油气井工程 [M]. 北京：石油工业出版社，2016.
- [24] 陆基孟，王永刚. 地震勘探原理 [M]. 青岛：中国石油大学出版社，2011.
- [25] 孙赞东，贾承造，李相方. 非常规油气勘探与开发 [M]. 北京：石油工业出版社，2011.
- [26] 李鸿乾，赵金海，彭军生. 石油钻井实用英语 [M]. 青岛：中国石油大学出版社，2006.
- [27] 张勇. 石油天然气钻井相关专业井控技术 [M]. 北京：石油工业出版社，2019.
- [28] 潘一. 钻井工程 [M]. 北京：中国石化出版社，2015.
- [29] 王明信，张宏奇，于曼. 油田地面工程基础知识 [M]. 北京：石油工业出版社，2017.
- [30] 吴世逵. 石油储运基础知识 [M]. 北京：中国石化出版社，2006.

石油专业核心词汇汉语拼音索引

A

API 油水分离器　198
安全系数　110
安全仪表系统　210
安装钻机　98
胺处理法　220
胺法工艺　222
胺法脱硫　221
胺溶液　220
胺吸收法　277
凹陷　8

B

拔顶油　292
白土精制　277
白云岩　3
板块构造学说　24
板式换热器　210
板式聚结器　203
半地下储罐　175
半干法烟气脱硫　278
半固定式系统　188
半固态　269
半潜式平台　90
伴生的碳氢化合物　152

伴生气　152
伴生气顶　65
饱和度　26
饱和式磁力仪　38
饱和压力　134
爆破员　52
爆炸　50
爆炸组　52
背斜　6
背斜圈闭　31
苯　301
泵站　246
边水驱动　133
边缘带　31
变质过程　17
变质岩　4
变质作用　17
便携式仪器　37
表层　99
表层套管　108
表观黏度　96
表面活性剂　94，242
表面活性剂驱　146
表面张力　94，146
丙烷　299
丙烯　300

并行塔 285
薄膜分离法 220
驳运 257
铂 280，304
捕集器 203
不分散钻井液 92
不混溶液体 190
不满负荷运行 234
不渗透的岩层 27
不透水层 155
不整合 29
部分混相驱替 145

C

采出气 212
采出水处理 193
采收率 143
采油 156
采油测试系统 168
采油树 118
采油指数 128
残余油饱和度 64
残渣燃料油 291
侧馏分 269
侧线 271
侧钻井 74
侧钻水平井 74
测井 119
测井车 120
测井电缆 123
测井队 77

测井评价 67
测井曲线 43，120
测井曲线图头 122
测井员 79
测网 50
层间干扰 127
层流 209
层流流态 240
层序地层学 10
层状结构 151
柴油 292
柴油发动机 80
产层 67
产层/目的层 112
产出流体 160
产量 136
产能评价 62
产能预测 67
产品混合 272
长链正构烷烃 302
常规油气藏 129
常规油气资源 15
常减压蒸馏装置 265
常压储气罐 175
常压汽提塔 221
常压容器 208
常压瓦斯油 270
常压蒸馏 264
常压蒸馏塔 265
场地使用权 97
超大型油轮/巨型油轮 256

超低渗储层　26
超级油轮　256
超声波脉冲　121
超重原油　159
车用机油　298
车用汽油　293
沉积构造　8
沉积模式　20
沉积盆地　8
沉积体系　9
沉积物　8
沉积相　20
沉积岩　2
沉降池　228
沉降罐　179
沉降器　284
成藏　16
成藏组合　16
成品油管道　231
成品油运输轮　255
成岩作用　17
成因解释　61
齿轮泵　186
冲积层　10
充气钻井液　92
重整反应冷却器　287
抽提装置　157
抽油泵　139
抽油杆泵　138
稠油开采　143
出油管/采气管　165

初次分离　192
初次运移　15
初次注水泥　111
初级产物　290
初级处理　194
初级分离器　198
初探井/勘探井　75
除草剂　309
除泥器　88
除气器　88
除砂器　88
除雾器　204，266
除雾设备　200
储采比　126
储层　27
储层空间　32
储层渗流　148
储层压力　67，127
储存设施　174
储罐泵　185
储量　68
储量丰度　69
储量评价　68
处理厂　165
处理设施　173
穿越　55
船上储存　186
纯油区　65
醇胺法除酸气　218
磁场　38
磁法勘探　38

磁化率　38
次生地层圈闭　29
次生孔隙　7
次生气顶　135
次要储集岩　28
丛式井　74
粗粒化　229
粗炼　262
醋酸纤维素　308
催化剂　279
催化剂降解　280
催化加氢裂化　284
催化裂化　279
催化裂化装置　281
催化脱蜡　284
催化重整　287
催化重整器　287
萃取用溶剂　273
萃余相　274

D

打捞　124
打捞篮　124
大地电磁场　39
大地电磁勘探　39
大地电流　40
大陆架　11
大陆坡　11
大鼠洞　98
大斜度井　74
大型油/气田　69

单纯形过滤器　206
单点系泊　257
单段压裂　116
单管道　237
单级分离　171
单向阀　182
单质硫　222
氮肥　309
氮气或二氧化碳气体压裂　116
导电性　41
导管　108
道密度　53
等厚线图　58
等时线/等时图　58
低产井　148
低固相钻井液　92
低剂量水合物抑制剂　242
低聚/齐聚反应　288
低硫原油　305
低渗透气藏　151
低渗透油藏　25
低温冷藏罐　180
低温下　242
低压储气罐　175
低压燃气管道　234
滴,细流;滴流,细细地流　268
底部沉淀物及水分　195
底部组分　279
底阀　132
底辟构造　19
底水　31

底水驱动　133
底水油藏　26
底拖法　235
地表溢流　155
地层测试　106
地层层序　10
地层储水能力　22
地层评价　64，119
地层圈闭　28
地层水　66
地层学　2
地层压裂　153
地层岩性油气藏　27
地层因素　66
地滚波　54
地壳海洋化　24
地面罐　175
地面油嘴　130
地面重力测量　37
地球化学勘探　34
地球物理测井　43
地球物理测井数据　119
地球物理电缆测井　43
地球物理勘探　34
地球物理异常　37
地下罐　175
地下资源　33
地形摄影　39
地震　161
地震波　55
地震测线　50

地震层序　47
地震成像　61
地震队　47
地震分辨率　60
地震勘探　46
地震剖面　58
地震剖面图　59
地震射线　54
地震数据　60
地震相　46
地震学　46
地震学家　46
地震仪　48
地震折射　57
地震子波　55
地质构造剖面　44
地质界面　58
地质勘探　34
地质录井数据　119
地质评价　62
地质图　44
点火处理 / 燃烧　147
点震源　49
点状或平均电阻式温度检测器　186
电测井　121
电磁场　39
电磁法勘探　40
电法勘探　40
电缆测井　120
电潜泵　139
电容　184

电位差 41

电子定位系统 42

电阻加热系统 241

电阻率 41

电阻率仪 41

吊钩 82

吊卡 86

丁二烯 301

丁烷 300

丁烯 301

丁烯烷基化反应 289

顶驱钻井 73

顶替液/驱替液 93

定阀 132

定向 107

定向井 74

定向取心 107

定向钻井 72

动态平衡吸水量 259

陡斜油藏 136

堵塞和弃井 107

端点放炮排列 51

短路 209

短时排放 230

段塞储集体积 201

段塞流 243

段塞流捕集器 200

断层 5

断层圈闭 31

断块背斜 6

断块油藏 25

断裂 24

断陷盆地 9

顿钻钻井 72

多层/多次完井 114

多次波 57

多次覆盖开关 53

多袋式过滤器 207

多底井 75

多分支井 74

多管封隔器 130

多管型段塞流捕集器 201

多管注水泥法 110

多级反应器 267

多级分离 196

多级分离工艺 196

多级吸收器 217

多级压裂 116

多级重整 287

多相流 190

多相流管道 237

多相流集输系统 173

多相流模拟程序 239

多组分特点 189

E

扼流压力阀 131

二层台 99

二次采油 136

二次处理 194

二次电磁场 40

二次混浆 112

二次完井　113
二次运移　15
二级井控　103
二甲苯　301

F

发电机　80
发泡剂　95
发散反射　56
发射器　48
发射线圈　123
反冲洗　112
反凝析气　213
反射波　56
反射波振幅　56
反射层　56
反射地震学　55
反射和透射系数　57
反射结构　56
反射路径　56
反射同相轴　59
反渗透法　229
反向断层　5
反演问题　57
反应器　283
反转构造　19
反转凝析　147
返排流体　117
返排期　153
返排设备　117
方井　97

方位变化率　105
方位角　105
方钻杆　82
方钻杆补心（方补心）　83
芳烃　302
芳烃浓缩物　275
芳香烃含量测试　275
芳香族烃　302
防腐蚀剂　95
防垢剂挤注技术　229
防喷器　89
防渗漏的储存池或储存罐　155
防水剂　307
放大器　49
放热反应　288
放射性测井　121
放射性同位素　122
放线员　52
非伴生气　152
非饱和烃　303
非常规天然气　150
非常规油气　15
非常规油气田　150
非常规资源　150
非挥发性溶质　273
非均质的　20
非均质性　7
非燃料产品　296
废弃化学药剂　242
废弃油砂　158
沸点　268

319

分布器　271

分层　55

分层层序　48

分馏　193

分馏厂　264

分馏塔　264

分馏装置　193

分配管道　166

分散体　192

分散钻井液　92

分选性　8

分液器　239

分支井　75

分子筛催化剂　279

分子筛法　222

分子筛网烘干机　202

酚醛树脂　304

风化层　10

风化作用　18

封隔器　114

扶正器　109

浮舱式顶　178

浮顶油罐　176

浮箍　109

浮盘　177

浮式生产储存卸货装置　171

浮鞋　109

浮选设备　207

浮油　227

浮子　185

浮子型油罐液位计　185

浮子液位计　183

辅助卡车　188

辅助油管　129

负压／欠平衡射孔　115

负载能力　259

复合圈闭　29

复合塔　272

富气　214

富气驱　145

G

伽马射线测井　122

改进原油采收率　143

钙处理钻井液　92

盖层　27

盖岩　27

感应测井　122

干井　76

干酪根　13

干气　141，213

干气回注　141

干气井　159

干式吸附工艺　224

干燥　263

刚性浮顶　178

港口　256

高 API 重度原油　190

高度变化　247

高沸点成分　267

高分子聚合物　309

高架储罐　175

高截频率滤波器　54
高能气体压裂　153
高品质芳焦　303
高闪点油　187
高温稳定剂　94
高辛烷值产品　294
高辛烷值汽油　294
高压储气罐　176
高压分离器　198
高压井口装置　118
高压配气管道系统　235
高压燃气管道　234
高压熔断器　210
高压水射流　286
高液位报警器　185
隔离液　111
隔油池　209
工业塑料　308
公共深度点　50
公用事业公司　170
拱顶罐　176
构造变形　19
构造分异　18
构造活动期　19
构造圈闭　28
构造图　45
构造—岩性圈闭　29
构造演化　18
构造应力　18
构造作用　18

古珊瑚礁　29
固定床催化裂化反应器　281
固定顶储罐　176
固定系统　187
固定油嘴　168
固井服务公司　77
固井装置　108
固井作业　108
固体超强酸催化剂　284
固体干燥剂脱水法　244
固体介质　224
固—液萃取　274
刮刀钻头　84
刮泥器　110
关闭装置　89
管道敷设　235
管道输气量　248
管道水动力分析　240
管道运输费　248
管道直径　236
管件　236
管式泵　139
管输油/净化油　231
管输油品分隔球　232
管线输送压力　247
罐底板　181
硅　280
辊子油（用于金属加工的润滑油）　297
过成熟烃源岩　21
过滤分离器　205

H

海底管道　233
海底管线　233
海底峡谷　9
海上浮式钻井船作业　73
海上可移动管道　233
海水钻井液　91
海相页岩　3
海洋地震勘探　46
海洋勘探　34
海洋运输　253
含硫化合物　241
含硫燃料系统　278
含气性评价　64
含气岩层　147
含水饱和度　26
含水层　31
含水醇类　215
含水量　274
含水率　64，243
含水区　194
含油饱和度　27
含油废水　226
含油量　64
含油气区带　32
含油区　64
焊接钢罐　181
航空磁测　39
航空电磁勘探法　40
航空勘探　34

航空汽油、航空涡轮汽油、航空涡轮煤油　294
航空燃油　294
合成材料（纤维、橡胶、塑料）　306
合成地震图　60
合成基钻井液　91
合成结晶硅—铝　280
合成气　295
合成树脂　309
合成药品　310
合成原油　158，294
核测井　121
核磁共振测井仪　120
核桃壳压力滤罐　226
珩磨技术　154
横剖面　44
喉道　8
后置液（压塞液）/尾液　111
候凝　112
呼吸阀　181
湖泊相　9
滑溜水压裂　116
滑轮组　81
滑套　131
化肥　306
化石　5
化石燃料　13
化探指标　66
化学精制　277
化学驱　144
化学溶剂法　275

化学试剂　310
化学添加剂　160
化学吸附过程　218
化学吸收法　218
化学抑制剂　241
环空流速　112
环烷烃　302
环形防喷器　89
环形浮舱　178
环形空间　100
环氧树脂　304
缓冲罐　179
缓冲液　111
换热器　271
挥发度　194
挥发性液体　177
回炼比　284
回流　270
回注产气层　225
混合产品　238
混合芳烃　275
混凝沉降罐　226
混凝土覆盖层　249
混相气驱　145
混相输送　237
混油　238
混油输送管道　238
活性炭吸附法　217
火成岩　4
火炬燃烧　225
火驱　159

火烧油层　144
货车运输　258

J

J 型铺管法　235
机械制冷　218
机械坐封封隔器　130
基本组分　263
基岩　4
基质孔缝　7
基质渗透率　151
激发极化　42
级间压力/门限压力　195
集气管线　167
集气站　173
集输管道　165
集输管网　164
集输系统　164
集输系统的输送能力　170
集油支线　166
计量分离器　171
计量设备　170
记录车　50
记录剖面　44
加工厂　172
加工热水器　286
加料泵　267
加密井　75
加气站　245
加氢裂化　282
加氢裂化原料　282

加氢裂化装置　282
加热储存　169
加热炉　272
加重材料　94
甲板污水排放　227
甲苯　301
甲醇　301
甲烷　299
尖灭　29
间歇供气　259
间歇气举　137
间歇式溶剂采收工艺　274
检波器　48
检波器站　48
检波器组合　48
减径管　236
减黏裂化　283
减压阀　167
减压瓦斯油　270
减压渣油　270
减压蒸馏　264
减压蒸馏塔　265
碱处理法　221
碱敏　127
碱水驱　145
碱洗　220，276
碱性溶液　223
浆体稳定性　95
降滤失剂　95
降凝剂　160
降斜　106

降液管　199
交货终点站　174
交接计量　184
胶结强度　110
胶结作用　17
胶黏剂　310
胶凝特性　96
焦化　285
焦炭分离　286
焦炭塔　286
焦油砂　157
角度不整合　24
绞车　81
绞车滚筒　82
搅拌　158
搅拌器　87
接单根　98
接近绝热的　258
接收器　48
节理　7，155
结垢　227
结晶　263
结晶器　275
结蜡过程　249
截断阀　236
截止阀　181
解堵　148
解释员　61
解吸作用　219
介电性　42
界面张力　24

金刚石型晶格　241

金刚石钻头　84

金属盐　305

紧急自动切断阀　168

进口分流器　199

进料　285

进料管　267

近海油气勘探　35

经济可采盆地　152

精加工设施　158

精炼　276

精馏塔　265

井场　97

井场监督　78

井底泵　169

井底回压　148

井底钻具组合　86

井架　80

井架工　77

井控设备　89

井口　89

井流　164

井滤泥饼　101

井漏　103

井喷　102

井侵　101

井筒地震技术　59

井网　126

井网密度　126

井下安全阀　132

井斜变化率　106

井斜方位　105

井眼轨迹　104

井眼清洁　106

井眼曲率　105

井涌　101

净产层　68

净气器　200

静电荷　176

静态平衡吸水量　259

静液面　140

镜质体反射率　66

纠斜　105

救援井/减压井　75

聚合物　155

聚合物驱　145

聚合物钻井液　92

聚合作用　283

聚晶金刚石复合片钻头　84

聚能射孔弹　115

聚酯纤维　304

卷管式铺管法　235

绝对渗透率　23

绝热管网　240

绝热容器　240

均质的　20

K

卡瓦　83

卡住的清管器　252

开采寿命　156

开发井　76

勘测员　36
勘探钻井　73
抗爆指数　289
抗压强度　110
苛性钠溶液　223，278
壳管式热交换器　210
可采储量　69
可控震源　49
可调油嘴　168
克劳斯硫回收法　223
空气预热器　275
空气钻井　72
孔隙　7
孔隙度　22
孔隙流体　30
孔隙性储油岩　26
孔隙压力　22
孔隙演化　22
库存罐　180
库存油品监测　183
跨接管　166
矿场自动接收、取样、计量、传输系统（LACT）　173
矿区／油田　169

L

蜡沉积　250
蜡分散剂　251
蜡晶体　250
蜡晶体改变剂　250
蜡泥　250

冷采　144
冷凝　263
冷凝器　268
冷却机　240
冷却水喷淋系统　207
冷却水系统　227
离开排列　51
离心泵　87
离心或旋风分离器　199
离心力　204
离心式压缩机　246
立根　99
立式分离器　197
立式隔油罐　208
沥青　290，299
沥青砂　14
连接／上扣　99
连续地震剖面法　58
连续气举　137
连续速度测井　43
炼油厂　262
炼油废水　226
梁式泵　139
梁式抽油　138
两级分离　196
两级过滤　216
两相分离器　197
两相流动　238
亮点技术　59
量水尺　182
量油尺　182

量油膏 206
裂缝/缝隙 7
裂缝形态 23
裂缝型孔隙介质 23
裂缝压力 23
裂化 278
裂解气 147
裂解塔 282
磷肥 309
零泄漏油库 187
流程选择 224
流动安全 169
流动形态 240
流化床催化裂化 281
流化床反应器 281
流量 166
流量控制技术 156
流量联轴器 171
流散火灾 188
流体焦化 286
流体界面 65
流体相 9
硫处理装置 277
硫回收 216
馏分 264
漏斗混合器 86
漏斗黏度 96
陆上管道 232
露点 195
露天开采 157
露头 5

螺杆泵 140
螺杆抽油泵 139
螺接库存罐 180
裸眼井 123
裸眼完井 114
落鱼 124
绿色完井/减少排放完井 160
氯乙烯 304
滤饼 101
滤介质 206
滤失控制 101
滤失量 101

M

马达法辛烷值 289
埋地管道 232
毛细管力 15
毛细管压力 127
煤层 154
煤层气 14，151
煤油 291
猛喷井 102
密度 36
密度测井 122
密封不漏蒸气 177
模拟熟化 65
母岩浆 4
目标岩层 153

N

耐磨接头 131

内壁腐蚀　249

内衬　98

内浮顶罐　177

内径　236

内涂层管道　165

逆流洗涤　278，285

黏度　268

黏合剂　307

凝聚过滤器　207

凝析　238

凝析气藏　26

凝析烃　239

凝析液　239

凝析油　147

凝析油段塞　201

扭转式超声分离器　199

P

PVC 聚氯乙烯　308

排放管桩　209

排涝泵站　186

排列　51

排气　147

排气口　275

排砂　208

排烃机制　20

排泄管　228

盘形顶　178

泡沫比例车　188

泡沫灭火系统　187

泡沫排放装置　188

泡沫排水　146

泡沫压裂　116

泡罩　272

炮点　50

炮井　53

配方　155

配气管网　235

配水管　172

喷发前兆井　102

喷射式钻头　85

批处理　224

漂缆技术　63

撇油器　229

贫气　214

贫油吸收法　218

平板照准仪　36

平底油船　254

平衡压力钻井　73

平均电位梯度　42

评价　62

评价井　76

破乳　191

破乳剂　192

铺管船　235

铺散器　199

普查勘探　34

Q

起爆药　310

起伏地形　247

起下钻　100

起钻　99

气/油接触面　65

气藏　25

气顶　31，135

气浮除油　203

气井排水　154

气举阀　138

气侵　102

气侵钻井液　92

气驱　135

气态　244

气体放空　208

气体分馏　269

气体覆盖层　204

气体脱硫　216

气体洗涤器　203

气相　9

气旋式系统　200

气压冷凝器　272

气页岩　151

气液比　170

气液发生器　217

气液分离器　190

气液两相流　189

气油比　200

气锥　136

弃井　76，161

汽化　270

汽轮机油　297

汽提塔　268

前导浆　111

浅变质岩地层　5

浅层过渡气　16

欠平衡压力钻井　73

桥塞　112

撬装设备　172

亲水的　192

亲烃的　158

亲油　192

亲油的　192

亲油化度　126

轻裂解　279

轻烃　158，267

轻烯烃　300

轻质汽油　190

轻质燃料油　293

轻质润滑油　297

轻致密油　160

氢键　220

清管器　251

清管器发射站　251

清管球　251

清洁剂　306

清蜡　149，250

清蜡剂　251

穹窿　6

球阀　182

球罐　180

球型分离器　197

区带评价　63

驱动方式　133

驱替液　141

驱替作用　133
驱油效率　148
取心　44
取样器　195
圈闭　28
圈闭机理　24
圈闭评价　63

R

燃点　187
燃料油/燃料　291
燃料油规格　291
燃气管网　234
燃气接收站　257
燃气输配系统　234
热采　143
热成熟度　21
热催化剂颗粒　280
热固性塑料　309
热活化钻井液乳液　93
热交换设备　210
热解反应　288
热力驱　144
热裂解　279
热水驱　144
热塑性塑料　308
热碳酸钾处理法　222
热洗/热油冲洗　209
人工产生的　41
人工举升　136
人工举升设备　137

人工源方法　35
人造纤维　308
日用化学品　310
容积表　181
容积式流量计　170
容器式段塞流捕集器　200
溶剂　296
溶剂采收　273
溶剂萃取　273
溶解气　135
溶解气驱　135
溶解油　227
溶解杂质　202
溶气气浮法　203
溶蚀作用　18
乳化剂　95，191
乳化物　191
乳化油　191
入口分流器　208
润滑剂　95
润滑油　297
润滑油基础油　296
润滑脂　297

S

赛璐珞　308
三级井控　103
三通　236
三维地震勘探　47
三维覆盖　47
三相分离器　197

三相混合物　237

三乙胺　303

杀虫剂/农药　306

砂岩　3

筛管/带眼衬管　114

闪蒸　185，202

闪蒸罐　266

扇形排列法地震勘探　49

上覆岩石　28

射孔　115

射流泵　140

深度值　58

深冷储罐　180

深水勘探　35

渗透率　23

生产层　129

生产分离器　196

生产井　76

生产设备　154

生产套管　109

生产完井　113

生成　13

生烃潜力　21

生物层序地层带　11

生物气　14

生油层/烃源岩　27

声波测井　43，121

湿法烟气脱硫　277

湿气　152，212

石化半成品　296

石化原料　299

石灰岩　3

石蜡　249

石蜡侧链断裂　283

石蜡油（凡士林）　297

石墨产品　298

石脑油　305

石油　12

石油储罐　174

石油地质储量　68

石油地质学　2

石油化工厂　290

石油化工行业　290

石油化学产品　306

石油焦　293

石油蜡　298

石油沥青防腐层　210

石油炼制　262

石油密度计　206

石油气（炼厂气或液化石油气）　290

石油重质尾部馏分　303

时差三维地震技术　47

时间跨度　21

实际泵排量　140

食品蜡纸盒　307

试井　67

试水膏　206

视平移/垂直截面位移　105

室内燃气管道　234

疏水阀　182

输气干线　167，231

输送损失　248

输送压力　247
输油泵　186
输油臂　237
数据采集　60
数据处理　60
数据解释　60
衰减作用　54
双层完井　113
双联过滤器　207
双盘结构　178
双筒式分离器　205
双闸板防喷器　90
水包油乳液　93，191
水合物　215
水合物抑制剂　241
水基钻井液　91
水解　220
水力旋流器　203
水力压裂　115
水龙带　82
水陆两用车　33
水敏　127
水泥评估　112
水泥造浆量　93
水平产层　152
水平管道　232
水平井　74
水平位移　105
水平钻井　72
水侵　102，134
水驱　133

水驱替　141
水溶相　30
水溶液　243
水头损失　209
水洗　219
水洗脱硫/加氢脱硫　276
水相　9
水油比　170
水蒸气汽提　282
水锥　134
司钻　78
四维延时成像　61
四乙基　304
速度流量计　258
塑料制品　308
塑性流动　20
塑性黏度　96
酸化处理　154
酸化压裂　154
酸气　213
酸气处理系统　285
酸性气体　213，273
酸性气体脱除工艺　223
随钻测井　122
随钻测量　107

T

塔顶冷凝器　221
塔顶气　213
塔盘　266
探边井　76

探测/勘探半径 43
探井 43
探明储量 69
探头 123
碳氢化合物 12
碳酸水驱 145
碳酸洗 219
碳酸盐岩 3
碳酸盐岩油藏 26
套管队 77
套管防喷法兰短节 131
套管井 123
套管内侧钻 104
套管配件 109
套管强度 110
套管射孔完井 114
套管特性 110
套管头 114
套管鞋 109
套管悬挂器 131
套管—油管环空/油套环形空间 130
套筒式防喷器 90
特高含水 148
提高采收率 143
提环 82
提升系统 81
天车 81
天然沥青 157
天然裂缝 6
天然气 12
天然气产能测试 146

天然气处理厂 212
天然气管道 232
天然气管道标识 249
天然气过滤器 215
天然气开采 146
天然气冷却装置 216
天然气离心式压缩机 245
天然气立管 201
天然气凝析液 159，214
天然气提纯 224
天然气脱水能力 243
天然气脱水器 215
天然气压缩站 246
天然气液化厂、气化厂 179
天然气液输送设施 233
天然气与液化气运输公司 260
天然汽油 190，239
添味剂 258
甜气 213
铁路油罐车 253
铁路运输 253
烃分子 283
烃类泄漏 186
烃露点 196
烃蒸气 270
同斜岩 4
透射层析成像 61
推靠式钻头 85
拖船 254
拖运业 254
脱丁烷塔 287

脱硫　195，276

脱氢环化反应　287

脱水醇类　215

脱水干燥　223

脱水塔　204

脱水装置　223

脱酸气方法　219

脱盐　194

脱盐原油　305

W

瓦斯油　292

外壁腐蚀　249

外浮顶罐　177

外加水套炉　237

外输管道　166

完井层段　113

烷基化　288

往复泵　87

往复式压缩机　245

往复式柱塞泵　139

微晶蜡　298

微生物采油　146

微生物勘探　34

微体化石　5

微重力测量技术　37

围堰区　186

桅杆式井架　81

尾浆　111

纬度效应　37

未经处理的井筒流体　189

未经处理的天然气　212

温室气体排放　161

文丘里管/文氏管　171

稳斜　106

稳压罐　205

涡街流量计　258

涡轮流量计　258

卧式分离器　197

卧式隔油罐　208

污泥泵　229

污水处理设备　226

污水井　228

污油罐　179

无杆泵　139

无机成因　13

无机盐　305

无硫原油　305

无水压裂液　117

物理分离　193，262

物理溶剂法　274

X

吸附法　217

吸附精制　277

吸湿能力　259

吸收法　244

吸收剂　216

吸收介质　216

吸收率　238

吸收器　217

吸收作用　217

希腊火药　307

析蜡点　250

烯烃　300

稀土金属结晶　280

隙间水　30

瞎炮　54

下套管　100

下套管工作　108

下钻　100

纤维球滤料　228

现场记录　59

现场开采法/原地提取　157

线性磁异常　39

相对年龄测定　21

相对渗透率　23

相对振幅　56

相消干扰　51

箱式气举　137

响应图　52

向斜　6

消泡剂　95

硝化作用　288

销售管道压力　248

销售用气态烃露点标准　225

小鼠洞　98

斜井/斜孔　104

携带颗粒　200

卸下/卸扣　98

卸油罐　178

卸载　244

辛烷值　289

新构造运动　19

信噪比　54

型号/级别　232

修井　123

修井工　79

修井工程师　79

修井公司　77

修井井口装置　124

修路或开路小组　33

絮凝　205

絮凝剂　227

悬浮固体　202

悬空侧钻技术　104

旋流分离器　198

旋塞阀　182

旋转分离器　204

旋转接头（水龙头）　82

旋转转向系统　85

旋转钻机　80

旋转钻井　72

选油站/油罐区　174

选择亲合性　274

循环注气　142

Y

压井　103

压力/真空阀　205

压力顶罐　176

压力缓冲器　246

压力控制阀　167

压力控制器　205

压裂树　117

压裂液　116

压气站　245

压实驱动　133

压实作用　17

压缩机　167

压缩机组　246

压缩天然气　214

牙轮钻头　85

岩石学　2

岩心分析　44

岩性　8

研究法辛烷值　289

盐构造　10

盐脊背斜　6

盐侵污染　228

盐丘　30

盐水　134

盐水钻井液　92

厌氧处理　228

遥感技术　35

遥控自动油罐计量系统　183

药深　54

野外地质调查　2

页岩层　151

页岩储层　10

页岩气　14，150

页岩油　156

液化处理厂　255

液化石油气　214，292

液化石油气船　257

液化石油气汽车罐车　253

液化天然气厂　212

液化天然气储罐　179

液化天然气船　255

液化天然气低温储罐　179

液化天然气管道　233

液化天然气罐车　254

液化天然气油轮　255

液量采收　168

液面控制器　204

液态　244

液体放空阀　204

液体干燥剂（乙二醇）脱水法　243

液体吸收方法　217

液位传送器　183

液位杆／液位尺　184

液位计量　183

液位开关　183

液位探测　184

液相氧化还原法　222

液压抽油机　138

液压电动挖土机　157

液—液萃取　273

液柱　184

一次采油　132

一次采油阶段　132

一次沉降　193

一次电磁场　40

一次反射　57

一级分离器　168

一级井控　103

移动床工艺　281
移动式系统　188
移动式钻井平台　90
乙醇　301
乙二醇回收装置　243
乙二醇脱水装置　215
乙烷　299
乙烯　300
异常物质　42
异丁烷　300
异构化　288
异构烷烃　302
抑制/冷却/凝析作用　196
抑制剂注入系统　242
阴极防腐　248
引爆机　52
隐蔽圈闭　29
硬沥青　299
硬质容器　181
油包水乳化液　93
油包水型乳状液　191
油驳　254
油藏　25
油藏监控　63
油藏特性描述　63
油藏压力　133
油藏增产措施　152
油层点火　144
油管　114，129
油管锚　130
油管头　130

油罐　174
油罐车　253
油罐出入管　185
油罐计量　183
油基压裂液　153
油基钻井液　91
油库　187
油路　260
油轮　255
油轮油舱　256
油泥　203
油品混输　232
油漆　310
油气比　170
油气储层评价　63
油气分离器　198
油气集输　164
油气检测　66
油气聚集带　32
油气勘探　33
油气苗　13
油气评价　62
油气田内管道　165
油气系统　32
油气显示　13
油溶相　30
油砂开采　156
油水过渡带　65
油水界面　65
油田工程师　79
油田管廊带　172

油田中心处理设施　169
油型气　16
油页岩　159
游车　81
游离气　134
游离水　191
游离水脱离器　197
游梁抽油机　138
有杆抽油　127
有机成因　12
有机物理溶剂　219
有利勘探区　33
有效孔隙度　22
有效烃源岩　21
有效吸水量　260
诱导气浮装置　229
预测储量　68
预热炉　266
原地　14
原动机　80
原溶剂　273
原生地层圈闭　28
原生孔隙　7
原生水　30
原始地层压力　126
原始溶解气油比　67
原油　12
原油干线　231
原油管道　231
远景圈闭　32
远洋油轮　255

运输管道　165
运移　15
运移路径　16
运载能力　257

Z

杂工　78
杂质　225
再沸器　271
再汽化　260
再生（解析）　218
再生反应　224
再生器　266，284
再生塔　221
造斜　103
造斜点　104
造斜率　106
造斜器　104
噪声道　51
增产　113
增黏剂　94
增压泵　246
增压站　247
渣油　292
闸板阀　181
闸板式防喷器　89
栅状图　42
炸药　52
照明煤油　291
折射速度　57
褶皱作用　17

真空泵 271	指进 134
真空闪蒸塔 266	指梁 99
真空脱水器 202	指向式钻头 85
真空蒸发 202	指型管 201
真空蒸馏塔 265	指重表 85
振动膜过滤系统 206	制动器 86
振动筛 87	制冷剂 307
震源 49	质子磁力仪 38
震源车 49	致密气 14,151
蒸发 263	致密砂岩 3
蒸发岩 4	致密岩层 159
蒸发逸失 259	致密油藏 25
蒸馏 263	智能清管器 251
蒸馏塔 265	滞留区 208
蒸馏装置 264	中间放炮排列 53
蒸气压 178	中间馏分油 298
蒸汽辅助重力泄油 142	中间石蜡 299
蒸汽喷射泵 272	中间套管 108
蒸汽喷射器 271	中心处理设施 128
蒸汽驱 144	中心控制台 185
蒸汽吞吐 142	中型油/气田 69
蒸汽压指标 193	中央处理站 173
正链烷烃 302	中央油库 166
正排流量计 195	终端产品 293
正弦波输入 52	重柴油 292
正压/过平衡射孔 115	重沸器 211
正演问题 57	重力场 36
正重力异常和负重力异常 38	重力沉降 199
支撑剂 117	重力分离 207
直井 73	重力分离器 198
直馏渣油 269	重力勘探 36

重力强度　36
重力泄油　136
重力仪　36
重馏分　285
重油（焦油）垫　14
重质馏分　269
重质燃料油　293
重质液体　268
重组分　269
轴流式压缩机　245
主补心　83
主要储层岩　27
主液体收集汇管箱　201
注氮气　141
注二氧化碳　142
注空气　142
注气　137
注入井　141
注水井　128
注水开发　140
注水泥塞　112
注烟道气　142
注油螺杆压缩机　209
注蒸汽　142
柱塞举升　137
转换工艺　222
转盘　83
转向马达　86
转运仓库　169
装货港　256
装药量　54

装油鹤管　172
浊流　10
资源评价　62
自动液位计　184
自喷采油　132
自然场方法　35
自然电位法　41
自升式钻塔　90
总产层　67
总孔隙度　22
总岩性　64
总重力分离　37
走滑断层　5
走时　55
阻火器　206
阻流管汇　102
组合　51
组合测井　122
组合货船　254
组合炮点　50
钻铤　83
钻杆　83
钻杆接头　83
钻工　78
钻工班　78
钻井泵　87
钻井队　76
钻井队长　77
钻井平台　90
钻井绳　81
钻井选址　97

钻井液　91

钻井液池　86

钻井液池检测仪　88

钻井液储备池　88

钻井液电阻率　94

钻井液静切力计　87

钻井液离心机　87

钻井液录井　120

钻井液录井图　120

钻井液录井员　79

钻井液密度　96

钻井液气体分离器　88

钻井液清洁器　87

钻井液循环系统　96

钻井作业　97

钻台　98

钻台工作队　78

钻头　84

钻屑　100

钻柱　84

最大含烃量　194

最大炮检距　53

最小炮检距　53

最终采收率　143